Math Basics FOR DUMMIES®

PORTABLE EDITION

by Charles Seiter, PhD

WILEY

Wiley Publishing, Inc.

Math Basics For Dummies®, Portable Edition

Published by
Wiley Publishing, Inc.
111 River St.
Hoboken, NJ 07030-5774
www.wiley.com

For general information on our other products and services, please contact our Customer Care Department within the U.S. at 800-762-2974, outside the U.S. at 317-572-3993, or fax 317-572-4002.

For technical support, please visit www.wiley.com/techsupport.

Wiley also publishes its books in a variety of electronic formats. Some content that appears in print may not be available in electronic books.

Library of Congress Control Number: 2006924000

ISBN-13: 978-0-470-05378-2

ISBN-10: 0-470-05378-X

Manufactured in the United States of America

10 9 8 7 6 5 4 3 2 1

1B/RR/QV/QW/IN

WILEY

About the Author

Charles Seiter wrote his first math book at the age of ten. It was a cartoon guide to calculus that was used as part of a junior-college math course. Since then, he has written 21 books on computers, math, and statistics.

Along the way, he got a PhD from Caltech but later abandoned academic life, with the help of TV game show money, and spent several years designing and teaching non-standard adult math classes in business math, statistics, and calculus. His message is simple: Anyone can learn everyday math with the right approach.

Publisher's Acknowledgments

We're proud of this book; please send us your comments through our Dummies online registration form located at www.dummies.com/register/.

Some of the people who helped bring this book to market include the following:

Acquisitions, Editorial, and Media Development

Project Editor: Tracy Barr

Acquisitions Editor: Kathy Cox

Editorial Program Coordinator: Hanna K. Scott

Editorial Manager: Michelle Hacker

Editorial Assistant: Erin Calligan, David Lutton

Cover Photos: © Corbis

Cartoons: Rich Tennant (www.the5thwave.com)

Composition Services

Project Coordinator: Kristie Rees

Layout and Graphics: Lauren Goddard, Denny Hager, Lynsey Osborn, Melanee Prendergast, Heather Ryan

Proofreaders: Leeann Harney, Evelyn Still

Indexer: Sherry Massey

Special Help: Carmen Krikorian

Publishing and Editorial for Consumer Dummies

Diane Graves Steele, Vice President and Publisher, Consumer Dummies

Joyce Pepple, Acquisitions Director, Consumer Dummies

Kristin A. Cocks, Product Development Director, Consumer Dummies

Michael Spring, Vice President and Publisher, Travel

Kelly Regan, Editorial Director, Travel

Publishing for Technology Dummies

Andy Cummings, Vice President and Publisher, Dummies Technology/General User

Composition Services

Gerry Fahey, Vice President of Production Services

Debbie Stailey, Director of Composition Services

Contents at a Glance

Table of Contents

Introduction

Plenty of people feel that they are "dummies" when it comes to math. You could say, in fact, that America suffers from math phobia or math avoidance on a massive scale. My modest ambition is simply to fix this situation with *Math Basics For Dummies,* Portable Edition. This book is intended to be quick and enjoyable in the aspects of math that you're probably going to encounter, have unfortunately forgotten, or are likely to need. Curious little facts about complex transcendental functions and other such things that might come in handy in a calculus course are missing.

Reading this book won't be like learning math in school. The emphasis here is on a relatively small number of examples, repeated in several variations so that you can find the one that's easiest for you to understand. There are no tests. I want you to get 100 percent of this material, so you won't see any trick questions, either.

Math Education and What's Wrong with It

On its own terms, math education in the U.S., despite decades of sincere efforts by platoons of dedicated teachers, is pretty clearly a failure. At the simplest level, we do a worse job than any other industrialized country. For more complicated reasons, results on standardized math tests inside the U.S. (not compared to other countries) have drifted downward for nearly 30 years.

Some people blame this trend on "new math" and other educational experiments. It's probably true that learning the binary number system never improved anyone's working arithmetic skills, but a careful survey of many years of test results shows that circumstances in the glorious old days of straight arithmetic drills weren't fabulous better.

I think the problem is that neither old-time arithmetic nor set theory are particularly interesting to most young students. Some kids actually like to memorize times tables, but most kids don't. If only the little rascals knew that set theory was setting their tiny feet on a path to a true appreciation of Lebesgue integral theory in an upper-division college math class, perhaps their hearts would beat faster.

Perhaps not. In this book, I propose to remedy the motivational problems of the traditional arithmetic class by telling you what teachers hope to accomplish in high school math. I can pick through all the material presented to you in 13 years of K–12 education (kindergarten to senior year in high school) and pull out 200 or so pages that are actually going to make a difference. As an added cultural literacy bonus, I'm also going to include a couple of issues that can cause you trouble in real life.

A Word about Calculators

Calculators are great. People were not meant to do compound interest problems using a blackboard and chalk, much less in their heads.

Yes, of course, it's sort of a disgrace to find yourself multiplying 2×10 on a calculator. But there's no need to feel ashamed at using one to multiply 17×395. When I was a university professor, my colleagues used to lament that replacing the slide rule with the calculator was the ruination of the student mind, and I've heard most of the other educational arguments against calculator use in classrooms.

These arguments are not only misguided in general terms, but they also don't apply here. As you look at the problems in this book, you're entitled to use any means necessary to get the answers you need. Getting the right answer, not purity of method, is the point.

Using This Book

The chapters in this book are free-standing and independent. Just find the chapter you need and start reading. If you need a brush-up in Algebra, go to Chapter 1. If Trig's your bag, go to Chapter 3. If you're now the proud user of a credit card and want to see exactly how much you're paying for the boots you charged, read Chapter 8. I've anticipated your every need, and I'm here to help.

Icons Used in This Book

A grand *For Dummies* tradition is the use of little icons to mark sections. Here's what the icons in this book mean:

TIP: This icon indicates a shortcut or an easier way to do a computation.

WARNING!: If a little inattention can lead to a wrong answer, or if you're likely to lose money easily, I'll flag it with a warning.

TECHNICAL STUFF: Can't be helped. Every now and then, I'll have to explain the basis of some operation (usually financial), and a little bit of algebra appears. You can skip almost everything marked with this icon, if you must.

REMEMBER: If it's important for you to remember a bit of information, I'll stick one of these icons next to it.

FUN FACT: Believe it or not, math has quite a bit of entertainment value, and lots of numbers have strange family histories. This icon points out some of that fun stuff.

EXAMPLE: You'll find practical examples of mathematical concepts marked with these icons.

The 5th Wave

By Rich Tennant

Chapter 1

What They Are Trying to Tell You in Algebra

In This Chapter

- A word about numbers
- The basics of notation
- Solving problems
- The simple facts about logarithms
- Higher math

Algebra in the high school sense is simply a set of rules for solving problems involving numbers. If you have trouble with algebra, the trouble probably comes from one of three sources:

- You have a hard time deciding which rule to apply to which problem.
- You have a hard time translating a problem from words into symbols.
- You have a hard time seeing the point of a bunch of apparently unconnected tricks.

Furthermore, algebra seems to be introduced to people at the exact point at which, for developmental, hormonal, or other reasons, their attention span for abstract symbolism has been reduced to a fraction of a second. Personally, I think that people should be taught algebra when they're 9 and social studies when they're 30. What's the point of learning about boring old Warren G. Harding in U.S. History if you're too young to hear exactly how he died in a hotel room in San Francisco? (Hint: He wasn't alone and apparently wasn't all that boring.)

I'm going to cover some highlights of the material that you may miss while your social life is evolving. It's actually pretty interesting, and a mere handful of tricks solves most of the problems that you encounter in real life.

The Numbers

For reference, I'm just going to list the kinds of numbers that underlie ordinary algebra first. Once I start talking about *x*s and *y*s, one of these kinds of numbers will get "plugged in" for the *x* or *y* value.

Integers

Integers are whole numbers — 3 or 17 or 365. "God made the positive integers; everything else is the work of man" is a quote from the famous German mathematician Leopold Kronecker. That means that 1, 2, 3, 4, and so on were already here in some sense, corresponding to the number of antelopes in a field or coconuts in a tree, before we even started to think about them. Right there at the end of our arms are little sets of integers that we can actually wiggle. Hey, they're even called digits.

The first human-invented integer was zero. There is reason to believe that zero first appeared as a notation for an empty space on a counting board (see Figure 1-1). When zero is used this way, it's possible to make a calculation on paper have the same form as the more familiar (in those times) calculation on a counting board.

We know who invented the lightbulb (all six inventors, actually), but we don't know who invented zero. It's known that it happened twice, independently, once in Central America for the Mayan counting system and once in India (some people claim that it may also have been invented in China at nearly the same time).

Finally, there is the man-made invention of negative integers. These numbers can be viewed as either an unnecessary idea for subtraction or, better, an extension of the idea of numbers lying along a line.

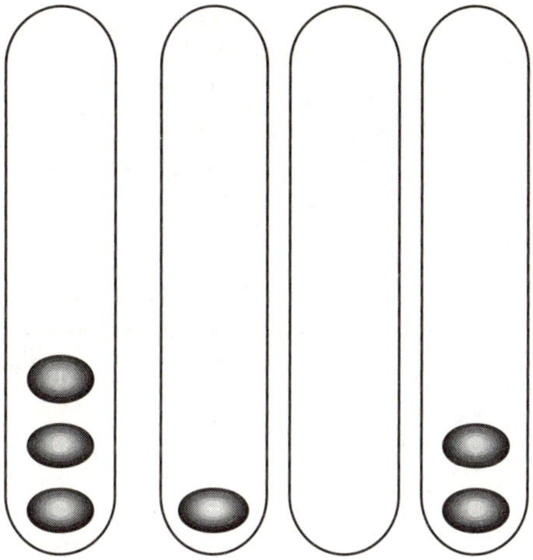

The number 3,102 on a counting board with beads. It could be represented as

III I · II

with the dot standing for "empty" or zero.

Figure 1-1: The origins of zero.

Take the number (–4), for example. You can add it to 6 to get the following:

$$6 + (-4) = 2$$

and you get the result 2. Why? Because adding a negative number is the same as subtracting a positive number. And what, you may ask, does subtracting a negative do? Well, as this example shows, subtracting a negative number is the same as adding a positive number:

$$6 - (-4) = 10$$

When you look at the numbers all strung out along the line in Figure 1-2, you see the rationale for the adding and subtracting of negative numbers. In the arithmetic operation, you are moving back and forth along the line. If you start at 2 and add 3, you move on out to 5. If you start at 2 and add (–3), the minus sign tells you to move to the left. Add means "move to the right" and subtract means "move to the left," but a negative sign on the number reverses this direction.

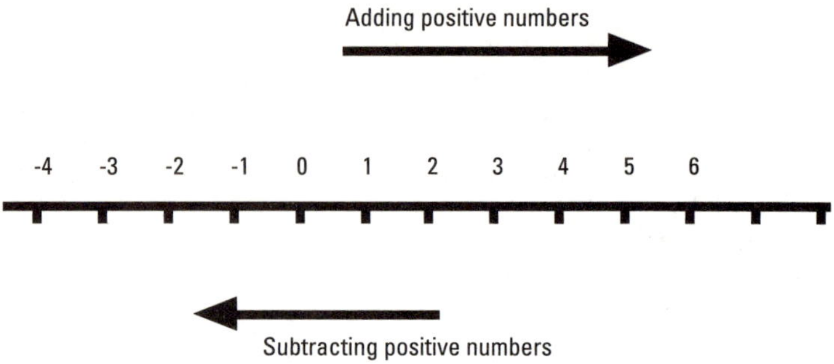

Figure 1-2: Numbers on a line.

Rational numbers

Rational numbers, which as a term include the whole numbers and all the fractions of whole numbers, are the next human invention — although referring to the previous example of antelopes on the plains, it should perhaps be noted that the concept of a half antelope was entirely familiar to lions millennia ago. In fact, the proverb "Half a 'lope is better than none" actually originated among lions.

Examples of rational numbers are

- $1/4$
- $3/8$
- $56/7$
- $59/7$
- $16/64$
- $355/113$

Just taking a look at these in order, only the first would have been considered a "decent" fraction in the ancient world. An Egyptian mathematical document called the Rhind Papyrus has a whole host of formulas for dealing with fractions, but all of them assume that the fractions will be expressed in a 1/something form.

For example, the next number would have routinely been written as

$$3/8 = 1/4 + 1/8$$

This fraction, in this exact peculiar usage, turns up later in history. When the Republic of Venice essentially hijacked the Fourth Crusade and directed it against Constantinople instead of Palestine, the Doge of Venice triumphantly proclaimed himself "Lord of a Quarter and Half a Quarter of the Roman Empire." That does actually sound more impressive than "Lord of 3/8 of the Roman Empire" or the even less compelling "Lord of 37.5 Percent of the Roman Empire, Discount Taken at the Register."

The next rational number in the list, 56/7, is not just a rational number; it's an integer, since

$$56/7 = 8$$

Just by taking this as a clue in a division problem, you can probably see that the next number in the list is a bit top-heavy, because

$$59/7 = 8 + 3/7$$

Another fraction that "needs work" is 16/64. By a lucky guess, you may be able to see that 16 divides evenly into 64. If you don't know this, you can still reduce the fraction to its basics by dividing the top (numerator) and bottom (denominator) by 2 to get

$$16/64 = \frac{(8 \times 2)}{(32 \times 2)} = 8/32$$

and then

$$8/32 = \frac{(4 \times 2)}{(16 \times 2)} = 4/16$$

and then

$$4/16 = \frac{(2 \times 2)}{(8 \times 2)} = 2/8$$

and then

$$2/8 = \frac{(1 \times 2)}{(4 \times 2)} = 1/4$$

What a complete waste of time! Fractions are still around because you can do so many little computational tricks with them, and because the publishers of books about sewing, cooking, and home improvement keep insisting on traditional measurements like $3\frac{1}{4}$ yards of fabric, $2\frac{1}{2}$ cups of flour, and $11\frac{1}{2}$ by $33\frac{3}{4}$ foot pieces of plywood. Until a couple years ago, even the stock market listed share prices as $423/8, a distant echo of "pieces of eight" and the origin of the term "two bits" for a quarter.

It's better, if the word better means anything at all, just to convert all this stuff to decimal fractions. In decimal fractions, $\frac{16}{64}$ becomes 0.25, which you can probably recognize right away as $\frac{1}{4}$, and if you can't recognize it, you can divide 1 by 4 on a calculator to remind yourself.

The last rational number shown here is $\frac{355}{113}$, a very good approximation for π. In real life, on computers and calculators and paper, you always have to settle for a rational approximation to special numbers like π, because neither you nor your computing device can deal with an infinite number of digits. How good is this particular approximation? If you use it to calculate the moon's orbit, you get the circumference of the orbit wrong by about the width of a fireplace.

Irrational numbers

These numbers can't be expressed exactly as fractions or as integers. One kind of irrational number can be expressed in the form

$$ab$$

where both a and b are rational numbers. The famous example of this expression is the square root of 2. For the square root of 2, *a* is 2 and *b* is $\frac{1}{2}$, a perfectly respectable rational number. By the definition of these things in algebra notation

$$\sqrt{2} = 2^{1/2}$$

This number starts out 1.41421356237309505... and continues forever. It's an irrational number, meaning that you can't express it just as a fraction. This class of irrational numbers can be constructed by a limited number of algebraic operations, though, since it has to be expressible in the *ab* format.

The truly weird kind of number is called *transcendental.* You can't put it into *ab* format, and, in fact, you can't even express it in a finite number of operations. The transcendental number that everyone knows is π, but another number

$$e = 2.71828182845904524...$$

turns up in a huge variety of calculus formulas. Most of those formulas are concerned with growth, which is why this strange transcendental number is at the heart of computations in compound interest. There's something pretty strange about calling on this spooky little number to calculate credit card interest.

A word about complex numbers

You don't see much about complex numbers in daily life unless you're an electrical engineer. But because you're likely to hear about them in algebra (or trig), here's the story. You already know something about square roots. You know that

$$\sqrt{4} = 2$$

and with a calculator you can show that

$$\sqrt{10} = 3.1623$$

Imaginary numbers provide a definition for the square root of a negative number. The number *i* is defined as

$$\sqrt{-1} = i$$

To show the kind of distinctions that I have spared you of thus far, I'll tell you that the preferred format of this definition is

$$i^2 = -1$$

Actually, it's considered better to say that the square root of –1 isn't really defined but that the square of i is. Similarly, the quantity $1/0$ is officially undefined, but in loose practical terms it's infinity (that's what your calculator will say, or overflow).

Infinity isn't really defined either, but as you divide 1 by smaller and smaller numbers, the result gets bigger and bigger.

One of the main applications of imaginary numbers is in defining a new number plane that has an astonishing range of properties and applications. Look at Figure 1-3. The *y*-direction in the usual Cartesian plane in geometry or graphing has been relabeled the *iy*-direction. The point (3,4), which used to stand for $x = 3$, $y = 4$, now indicates $x = 3$, $iy = 4$. All the numbers in this plane can be represented in the form

$$a + bi$$

Multiplying imaginary numbers in this form is not particularly hard. Look at this example:

$$(2 + i)(3 + 2.5i)$$
$$= 2 \times 3 + i \times 3 + 2 \times 2.5i + 2.5 \times i \times i$$
$$= 6 + 3i + 5i + 2.5\,(i2)$$
$$= 6 + 8i + 2.5\,(-1)$$
$$= 3.5 + 8i$$

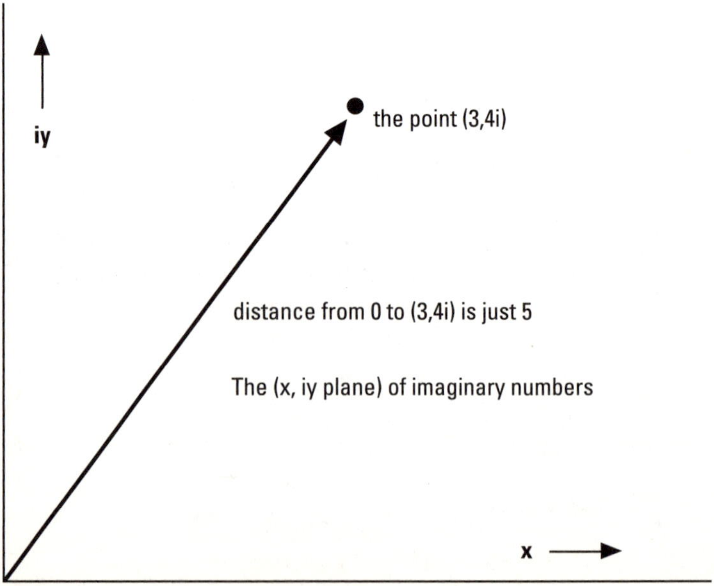

Figure 1-3: The imaginary number plane.

When complex numbers are expressed in another form, involving distance in the plane from zero and angle from the x-axis, it's even easier to multiply them. That, however, is another topic pretty far from everyday math.

The Rules of Algebra

One of the nice things about algebra is that you already know a lot of it, in the sense that most of what you expect to be true of numbers is also true of numbers represented by symbols.

Using letters to stand for any of the classes of numbers listed in the preceding sections, you have a small set of rules:

$$a + b = b + a$$

$$a + (b + c) = (a + b) + c$$

$$a \times b = b \times a$$

$$a \times (b \times c) = (a \times b) \times c$$

$$a \times (b + c) = a \times b + a \times c$$

Other algebras

The reason that textbooks go on at great length about these rules for real numbers is that they aren't necessarily true for other potential types of algebra. Mathematicians can define an algebra with special objects A and B and an operator +, where, for example

$A + B$ is not equal to $B + A$.

A lot of the most important mathematics of the 20th century concerns types of algebras in which the familiar rules that apply to real numbers just don't apply at all. It may sound strange, but these rather other-worldly algebras are important in many practical applications in physics. Elementary particles don't care what the story is on real numbers — they've got their own rules.

You can look at these equations with numbers plugged in for letters to convince yourself that for the ordinary numbers you can display on a calculator, these rules are always correct.

$$4 + 5 = 5 + 4$$

$$1 + (2 + 3) = (1 + 2) + 3$$

$$3 \times 7 = 7 \times 3$$

$$3.14 \times (5.3 \times 2.7) = (3.14 \times 5.3) \times 2.7$$

$$52 \times (34 + 13) = 52 \times 34 + 52 \times 13$$

For the purposes of this chapter, the rules that I just listed mean things as self-evident as

$$4 \times 5 = 5 \times 4$$

which you probably haven't had much doubt about lately. You probably figure that $2 + 3 = 3 + 2$ also. Because the algebra in this chapter concerns only real numbers, anything on both sides of an equation is a real number. That means that you can do both of the following:

- Multiply both sides of an equation by anything you like.
- Add or subtract anything you like to both sides of an equation.

If you are confronted with an expression like this:

$$4x^2 + 3x = 12x$$

you can

1. **Subtract $3x$ from both sides, giving $4x^2 = 9x$.**
2. **Divide by x on both sides, giving $4x = 9$.**
3. **Divide by 4 on both sides, giving $x = 9/4 = 2\frac{1}{4}$.**

(Actually, step two is a bit tricky. Another solution of the original equation is $x = 0$. It's always worth doing a fast check to see if $x = 0$ is a solution.)

What you're usually trying to do in algebra is to get to the definition of some unknown quantity in terms of a simple number.

By the way, a notation in which individual letters stand for the unknown quantities took a long time to evolve, and the now-familiar notation

$$x^2 \text{ for “} x \text{ times } x\text{”}$$

came later still. The *x*, *y* notation for unknown quantities was invented in the late 16th century in France by François Viete — one of the first people to write an equation that you or I would recognize as algebra. Before that, it was pretty much standard in European and Arabic math to write equations that were the equivalent of statements like

$$\text{Profit} = \text{Revenues} - \text{Cost}$$

keeping whole words as names for the quantities to be determined.

Problems, Problems

Most of the real-life applications of algebra consist of converting problems in words into problems in symbols, since it's usually easier to solve the symbol version of the problem. The emphasis on this aspect of algebra is simple to explain: You are very seldom confronted with quadratic equations (equations that have x^2 as the highest power of x in the equation), but a simple bit of algebra crops up almost every day.

Problem 1: Getting there

At 70 miles per hour, your car gets 25 miles per gallon. At 55 miles per hour, your car gets 30 miles per gallon. You are 270 miles away from Las Vegas on an empty stretch of desert road with no gas stations along the way. From long experience with your car, you can tell from the needle on the gas gauge that you have ten gallons left. It's two in the afternoon. When do you get to Las Vegas?

The first thing to do in this type of problem is separate it into cases. Take case 1 first, the 70 mile per hour case. The real question here is this: If you get 25 miles per gallon, how far can you go on ten gallons? From practical experience, you can say that on the first gallon you go 25 miles, on the next gallon you go another 25, and so forth until you've used up all ten gallons. In other words

$$\text{miles you go} = 25\ (\text{miles/gallon}) \times 10\ \text{gallons}$$
$$= 250\ \text{miles}$$

The gallons in this multiplication cancel (you have gallons on the bottom of one units-fraction times gallons in the other term), just as if they were *xs* or *as* or *bs*. In notation, you can say that the formula is just

$$D = m \times g$$

where

D = distance you go in miles

m = mileage in miles/gallon

g = available gallons

What you see from this demonstration is that if you start at two in the afternoon and drive at 70 miles per hour toward Las Vegas, you'll get there only after hitchhiking the last 20 miles. That could be a long time.

So apply the formula to the second case. If you drive more slowly, you get 30 miles per gallon as your mileage. (Actually, these numbers are fairly realistic.)

$$D = m \times g$$
$$300\ \text{miles} = 30\ \text{miles/gallon} \times 10\ \text{gallons}$$

You can make it all the way to Las Vegas if you drive more slowly. That realization gets you to part two of the question. At 55 miles per hour, how long will it take you to go 270 miles? The right formula is

$$\text{distance} = \text{miles per hour} \times \text{hours}$$
$$D = s \times t$$

where

D = distance in miles

s = speed in miles per hour

t = time in hours

What you want is a formula that says that t = something because you're looking for the time. But this formula will do, since you're left with a way to get t if you put in the two available numbers.

Plugging in the numbers, you get

$$270 \text{ miles} = 55 \text{ miles per hour} \times t$$

Multiply both sides of this by hours, divide by miles to clear the units, and the equation reads

$$270 \text{ hours} = 55 \times t$$

If you divide by the number 55, you get

$$4.91 \text{ hours} = t$$

It's going to take you almost 5 hours, as 4.91 is nearly 5. In other words, 4.91 hours is 4 hours plus

$$0.91 \text{ hours} \times 60 \text{ minutes/hour} = 54 \text{ minutes}$$

Looking at the mileage figures, you see that you risk running out of gas a few miles short of the bright lights of the fabulous Strip if you try to go any faster than this.

Problem 2: The ridiculous age problem

You see problems like this in high school, and you see problems like this on the SAT or ACT test. And then you never see them again, except perhaps in puzzles. You probably can have a very successful career in business without being able to solve this kind of problem at all. But because this part of the book aims to explain what they try to tell you in high school, I feel obliged to recapitulate.

George is four years less than three times Steve's age. Steve is now 12. How old is George?

Nothing like the experience of looking at the clock, seeing that you have three minutes left to complete an algebra test, and being confronted with this sort of question. Ever meet someone who claimed to be your age times the square root of π plus two? Anyway, to put it into symbols, just say that

S = Steve's age

G = George's age

Three times Steve's age minus four years gives you George's age. That is, George would be exactly three times Steve's age, but the problem says that he's four years younger than that. In symbols, that's

$$3S - 4 = G$$

Taking the multiplication statement first, if you plug in the number 12 for S, you get

$$3 \times 12 - 4 = 36 - 4 = 32 = G$$

George is 32.

Actually, problems like this are just exercises in translating a set of conditions into algebraic statements for solution. Often the problems are somewhat contrived, but practicing symbol-to-word translations is more than half the battle in algebra. The rest of the battle is just practicing all the "subtract from both sides, multiply by both sides" stuff.

Problem 3: The almost-ridiculous mixture problem

Here's another classic problem, repeated down through the decades in textbook after textbook. Around 1980, some algebra textbooks started featuring lavish use of color printing, but no one seems to have thought of updating all the old problems.

A five-pound can of mixed nuts contains cashews and peanuts. It costs $30. If peanuts cost $4 a pound and cashews cost $9 a pound, what is the amount of each in the mixture?

In this case, it's almost simpler to translate everything in the problem into notation right away than to try to think about the problem in these terms.

First, you can say that the number of pounds of peanuts in the can is

$$\text{number of pounds of peanuts} = p$$

since you don't know what the number is and the letter p is pleasantly evocative of peanuts. Having done so, you have also established the number of pounds of cashews. No, it isn't c. It's

$$\text{number of pounds of cashews} = 5 - p$$

because the problem states that it's a five-pound can. Everything in the can has to add up to five pounds, so

$$\text{peanuts} + \text{cashews} = 5$$

$$p + \text{cashews} = 5$$

$$\text{cashews} = 5 - p$$

The next expression to translate is the total cost information.

Thirty dollars has to come from p pounds of peanuts at $4 per pound and $(5 - p)$ pounds of cashews at $9 per pound. Gaily disregarding the units of pounds and dollars for the moment, you can write

$$30 = p \times 4 + (5 - p) \times 9$$

or, a bit more cleanly, you multiply 5 by 9 and $-p$ by 9 to get:

$$30 = 4p + 45 - 9p$$

Resorting to mere algebra, you add $4p$ and $-9p$ to get

$$30 = -5p + 45$$

Subtracting 45 from both sides gives you

$$-15 = -5p$$

Multiplying both sides by –1 gives you

$$15 = 5p$$

Finally, you divide both sides by 5:

$$15/5 = p$$

which, reduced, is

$$3 = p$$

The can contains three pounds of peanuts, which in turn means that it contains two pounds of cashews.

TIP

As a quick common-sense check, look at the numbers involved. The average cost of $30 worth of stuff in a five-pound can is $6 a pound (it's just $30 divided by 5). The peanuts are $4 a pound, and the cashews are $9 a pound. Because 4 is closer to 6 than 9 is, you can reasonably expect to see more peanuts in the mix than cashews. So the answer makes sense, which is always nice. Sometimes these problems are pointlessly arbitrary, but you can usually use a little common sense to check the answers.

Logarithms

I'm going to shift gears for a moment and talk about another use for algebraic notation. Suppose I formulate another problem in *x* that requires a bit more effort than plain arithmetic. Here it is:

$$8 = 2^x$$

You can poke around, try some numbers, and convince yourself that *x* can equal 3 since

$$2^3 = 2 \times 2 \times 2 = 8$$

And 2^4 (just putting one more 2 in the stack) is 16.

Now, in the definition of powers (or exponents) of numbers, a square root counts as a power of ½, or equivalently 0.5. Looking at this result

$$\sqrt{2} = 2^{1/2} = 2^{0.5} = 1.414$$

many pioneers in algebra came to an interesting conclusion. They realized that if you can use fractional exponents, you can generate any positive number you want as exponents of a single starting number, called a base.

For example, using the number 2 as a base, you can find 2 to the exponent 3.5 as

$$2^{3.5} = 2^3 \times 2^{0.5} = 8 \times 1.414 = 11.31$$

and other numbers between 8 and 16 can be generated by picking other fractional exponents (exponents between 0 and 1), it follows that you can generate an approximation of any number as a power of 2. In other words, the equation

$$N = 2^x$$

has a solution x for any positive real number N that you care to pick.

Table 1-1 gives you a little table fragment for use with this argument:

Table 1-1 World's Smallest Log Table

Log Base 2	*Number*
2.0	4.0
2.5	5.66
3.0	8.0
3.5	11.31
4.0	16.0
4.5	22.63

In this case, I've generated a table of powers based on the number two. The table, for historical reasons, is called a table of logarithms. In real life, logarithms using as a base the number 10 and the transcendental number *e* are the only ones that we typically use. Base 10 and base *e* correspond to the keys *log* and *in* on a calculator, respectively.

The reason this is important is that in pre-calculator days, adding numbers was seriously easier than multiplying them. If you make up a table of numbers and their equivalents as powers of some number (the base), then you can multiply these numbers just by adding the powers and looking back to the table for the original number that corresponds to your new power. In this table, if you want to multiply 4.0 by 5.66, you look up the logs (2 and 2.5), add them (you get 4.5), and read off the product from the table (4.5 corresponds to 22.63). The slide rules that engineers and scientists used until the appearance of calculators were a sort of logarithm table engraved on sliding strips.

You don't need to know much about logarithm tables anymore — the other day, I saw a Sanyo scientific calculator with log/ln keys in a drug store for $6. Of course, I bought it to add to my collection of about 20 completely unnecessary calculators. (Hey, it's cheaper than collecting 35mm zoom lenses.) As an additional curiosity, even the cheapest plastic slide rules of the 1960s and earlier were so thoroughly discarded when scientific calculators were developed that they are now expensive collectibles, like rare models of Barbie dolls.

Just as a demonstration, I'll show you the meaning of a couple of logs. Using a calculator, I find that

$$\log 115 = 2.0607$$

$$\log 12{,}400 = 4.0934$$

Log 100 would be 2 since $100 = 10^2$ (that's essentially the definition for log). Because 115 is a bit more, its log is a bit larger. This result also says that 10 to the 0.0607 power should be the factor 1.15. You can check this yourself on a calculator with a 10*x* key — and it checks, all right.

The log of 12,400 is just the same thing, scaled up. The log function essentially counts the number of decimal places and as the decimal part calculates another number that's a sort of correction factor to place the number approximately within the next factor of ten.

There's not much more to logarithms than that. They're used in science extensively for dealing with large numbers — after all, the log of a million is six, so they're good for scaling things down to a manageable size. But now that experimenting with logs on a calculator is quick and easy, a lot of the mystery is gone.

Higher Math

In algebra, the translation of a word problem often becomes an expression in *x*. If the final expression contains just *x* and numbers, you can always solve it. In fact, if an expression contains just one unknown (*x* is the unknown here) to a power less than five, you can always find solutions for *x*. But if the expression ends up containing higher powers of *x*, sometimes you can solve it and sometimes you can't. In fact, one of the main challenges facing mathematicians for centuries was to determine when you can and when you can't solve an algebraic equation.

In most day-to-day algebra problems, the highest power you encounter is a mere 2. Expressions containing x^2, called quadratic expressions, are not only especially common but also especially easy to handle. In fact, you can write a general solution for any quadratic equation. That solution is shown in Figure 1-4.

There's also a pretty complicated solution of equations that contain x^3, although you seldom need them. The reality is that, for finding the values of *x* that satisfy a general expression like this:

$$ax^5 + bx^4 + cx^3 + dx^2 + ex + f = 0$$

you usually need to have actual numbers plugged in for the *a*, *b*, *c*, and so forth, and you use some computer program to find the *x*s that work by a process that's actually a sort of enlightened trial-and-error method.

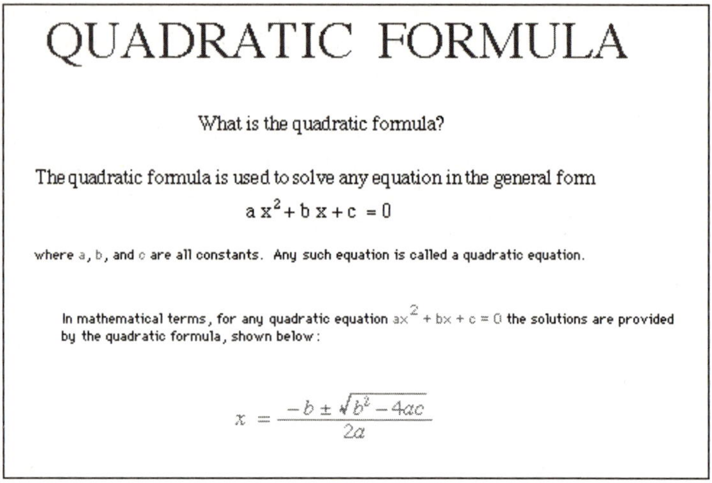

Figure 1-4: Quadratic expressions, automated.

Modern computer programs for mathematics typically have an automated method for finding the zeros of polynomials (these expressions in *x* and powers of *x* are all called polynomials, and you can also have polynomials in several variables). Figure 1-5 shows such a program finding approximate solutions to a nice, big equation.

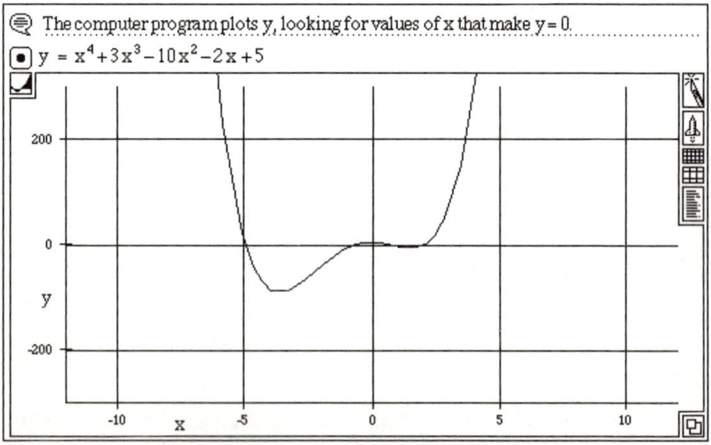

Figure 1-5: A computer solution of an equation.

If you have any interest in seeing how modern computer programs have taken over traditional topics in algebra, you might want to look at the program Theorist (from Maple Waterloo Software), available for Windows PCs and Macintoshes. Just as the program Quicken (see Chapter 9) can do anything you ever needed done for your checkbook, Theorist can solve every problem you ever face in high-school or college math. Although the scientists in Far Side cartoons still work at blackboards, in real offices and labs the work is done on a computer. Computer algebra has simply revolutionized equation-solving, probably forever.

Chapter 2

What They Are Trying to Tell You in Geometry

In This Chapter

- Geometry before Euclid
- The Greek ideal emerges
- Descartes changes it all
- Graphing makes geometry easier

Even now, a mere 2,200 years or so after Euclid, Euclidean geometry is considered to provide such an excellent mental exercise — often a student's first introduction to the idea of rigorous proof — that it remains part of the high school curriculum. What Euclid introduced into the large body of geometrical results available to the ancient world was the idea of the axiom. Instead of accepting the idea that geometry consisted of hundreds of unrelated results, rules-of-thumb, and "guesstimates," Euclid decided to find the smallest number of *axioms,* or fundamental principles, from which all the other rules could be derived.

There is a great deal to be said for this approach, especially for students who will be going on to advanced studies in math or science. It has the unpleasant side effect, however, of producing vast platoons of ex-students who remember only that they hated geometry and can't remember the formula for the area of a circle.

Nevertheless even though there aren't any shortcuts to mastering Euclidean proofs learning a few practical geometric facts is no great effort.

B.E. (Before Euclid)

The Babylonians and Egyptians, being rather more practical than the abstraction-minded Greek philosophers, assembled rule books covering most common land-measurement or volume-measurement situations. Learning these rules was part of the education of a scribe. Here are some rules that may be familiar to you.

Area of a rectangle

One of the earliest maps in existence is a little clay block measuring about six inches by three inches (the scribes preferred palm-sized clay tablets) from Mesopotamia. On it, you can clearly see a bunch of standardized riverfront lots going from a fixed boundary down to the Euphrates (see Figure 2-1). Later clay maps (there's a great city map of the city of Nippur from 1500 B.C.) showed the clarity of a modern subway directory. The Sumerians and their successors weren't much for theory, but even they could translate the imprecise boundaries of the land-map plots into ideal rectangles and calculate the area (see Figure 2-2).

Because plots of land were (and are) worth money, real areas were calculated directly on the ground rather than from figures in clay. And the favored measuring tool of the ancient Near East was the knotted rope, as shown in Figure 2-3. We actually have some real examples from the dry climate of Egypt (hemp rope doesn't keep well through the 5,000 years of humidity that constitutes the climate history of Sumeria) and can thereby reconstruct all sorts of geometric tricks of the ancients.

The first and simplest is the area of a rectangle. Take two ropes, line them up at the edges of the rectangle, and count all the little squares, as Figure 2-4 demonstrates. One side of the rectangle is three knots long, and one side is four knots long. If you count the internal squares, you have 12 of them. After you do this for a while, it becomes clear that multiplying the numbers for the lengths of the two sides is exactly the same operation as counting all the squares in the rectangle.

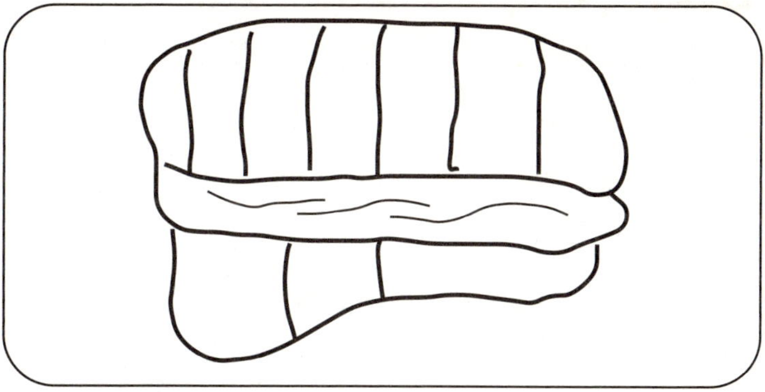

Sumerian Riverfront lots, depicted on a little hand-held clay map

Figure 2-1: The dawn of geometry: riverfront lots in Mesopotamia.

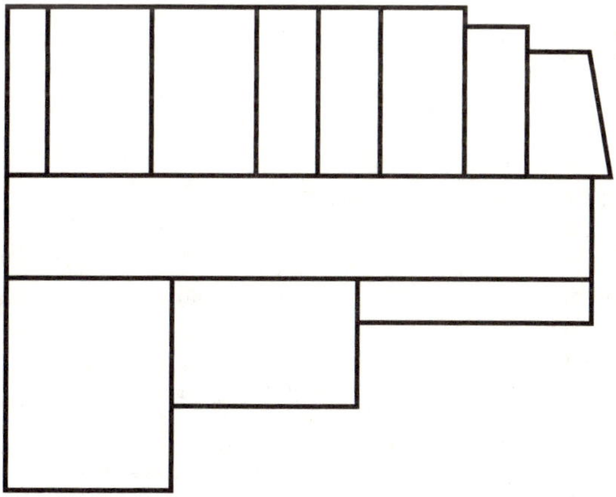

Sumerian Riverfront lots, reduced to straight-line geometry

Figure 2-2: The beginnings of abstraction: translating uneven plots into ideal rectangles.

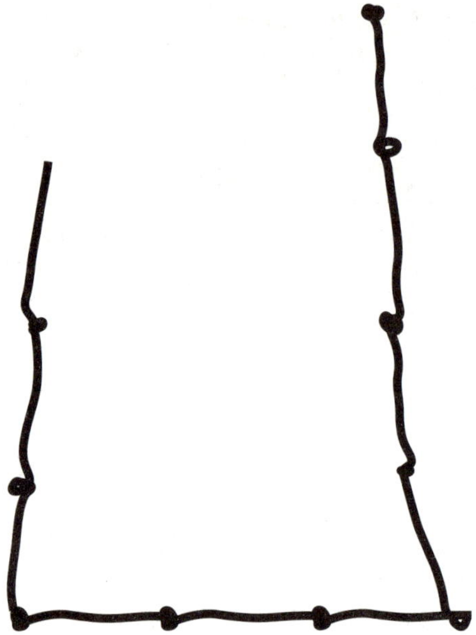

Figure 2-3: The knotted rope, the basic tool of geometry.

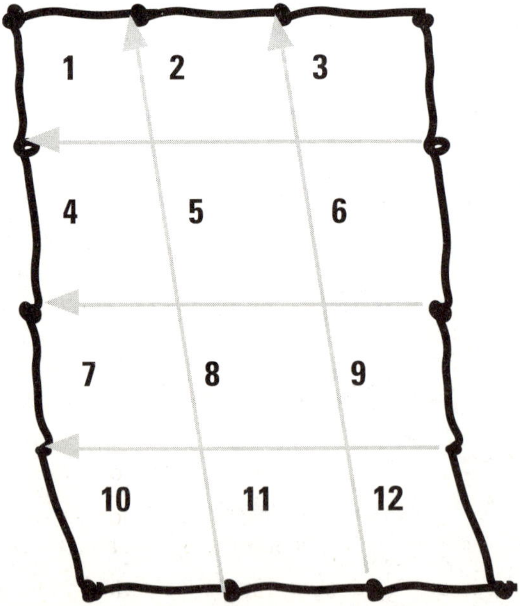

Figure 2-4: Multiplying and counting give you the same answer.

By the way, you can do the same thing if both sides of the rectangle turn out to have the same length, which is the special case of the square.

Area of a bent rectangle

Once you start laying out real riverbank plots on a hot riverbank in a summer in Sumeria in 2750 B.C., what do you think are the odds that the plots will be perfect rectangles? Imperfection is almost guaranteed. The real plots are probably going to look something like those shown in Figure 2-5.

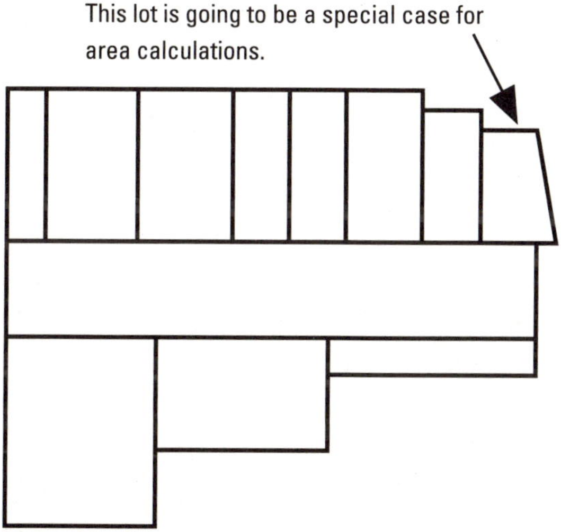

Figure 2-5: Many "rectangles" are somewhat less than rectangular.

What did the scribes make of this situation? They looked at it and made the following conclusions:

1. We have to multiply the long side by something.
2. If we multiply the long side by the shorter end near the road, the answer will be too small for the correct area.
3. If we multiply the long side by the longer end near the river, we are clearly going to get an area that's too large.
4. What the heck. Propitiate the goddess Inanna and split the difference. We'll take a number in between the

values of the two ends and multiply that by the long side. (See Figure 2-6.)

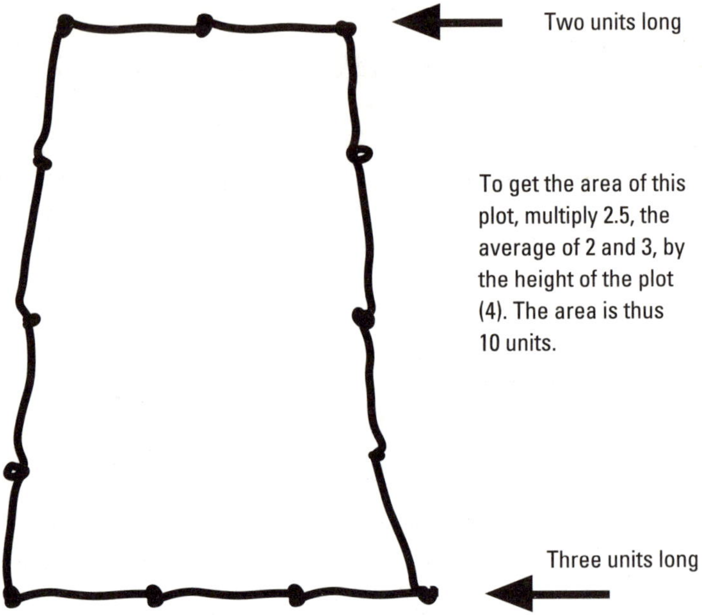

Figure 2-6: A trickier calculation using a value in between the values of the shorter and longer ends.

When the Greeks got around to developing Greek names for everyone else's geometrical objects 2,000 years later, they named this kind of rectangle a trapezoid. Please remember that at the time these calculations were going on in Sumeria, the Greeks were an illiterate Indo-European tribe with nothing to offer their neighbors but an inexhaustible supply of petty skirmishes.

Area of a triangle

This last result is actually fairly sophisticated because it leads the way to other area calculations. Look at the series of "rectangles" in Figure 2-7. The last one has ceased to be a rectangle at all and is now just a triangle. But the rule for calculating the area is still the same as it was in the "bent" rectangle case.

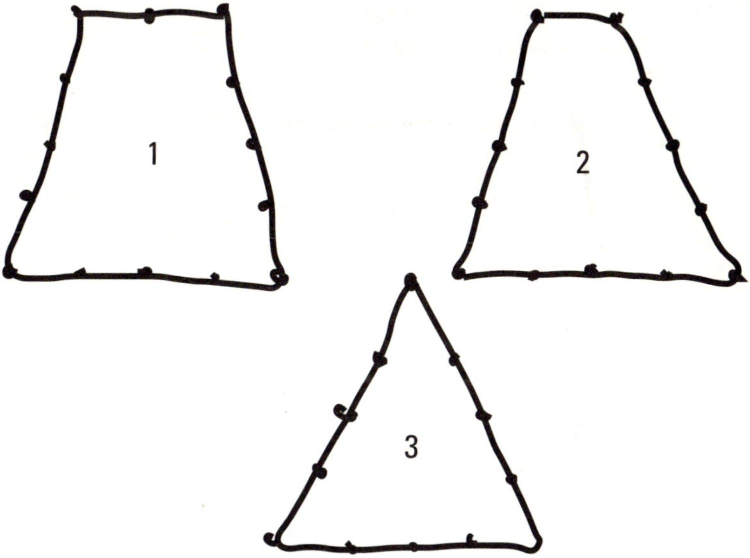

Figure 2-7: A series of trapezoids.

When the length of the short end drops to zero (that's the case of the triangle), the rule becomes

$$\text{area} = \text{length} \times \frac{(\text{short side} + \text{long side})}{2}$$

$$\text{area} = \text{length} \times \frac{(\text{side that's left})}{2}$$

In standard notation, this is the rule

$$A = \frac{(a \times b)}{2}$$

where the letters stand for the lengths in the diagram in Figure 2-8. This is the standard formula for the area of a triangle, usually derived by drawing a triangle inside a rectangle (see Figure 2-9) and comparing the areas of the different sections.

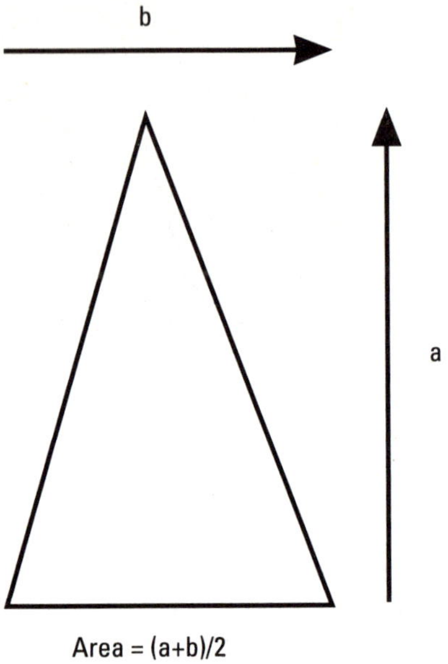

Figure 2-8: The triangle in standard geometry.

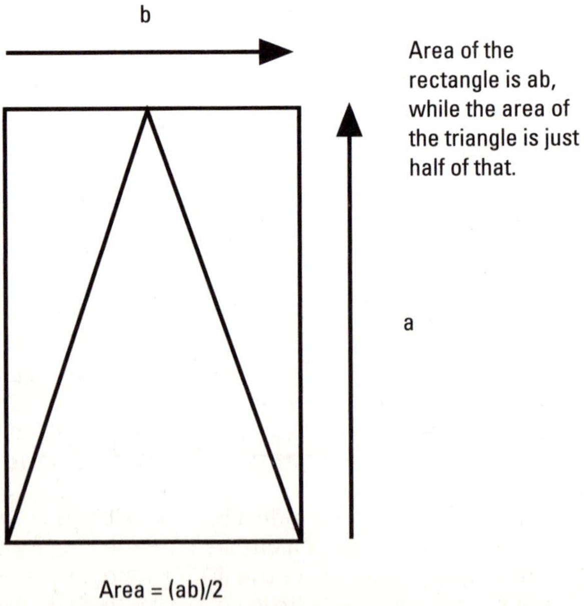

Figure 2-9: The usual derivation of the triangle area formula.

Area of a circle

Equipped with the formulas for the area of a rectangle and a triangle, geometers around 2000 B.C. were ready to measure the area of any sort of geometrical object, no matter how oddly shaped. Doing so is simply a matter of breaking up the object into rectangles and triangles, working out the area for all these little bits, and adding up the areas of the bits. As a practical consideration, a city official could work out the exact taxes on any plot of land, undoubtedly a comforting fact for the hard-pressed Sumerian peasant. No wonder these people invented beer.

But the method of calculating areas by looking at areas of rectangles or triangles is surprisingly powerful. Look at the two possibilities in Figure 2-10.

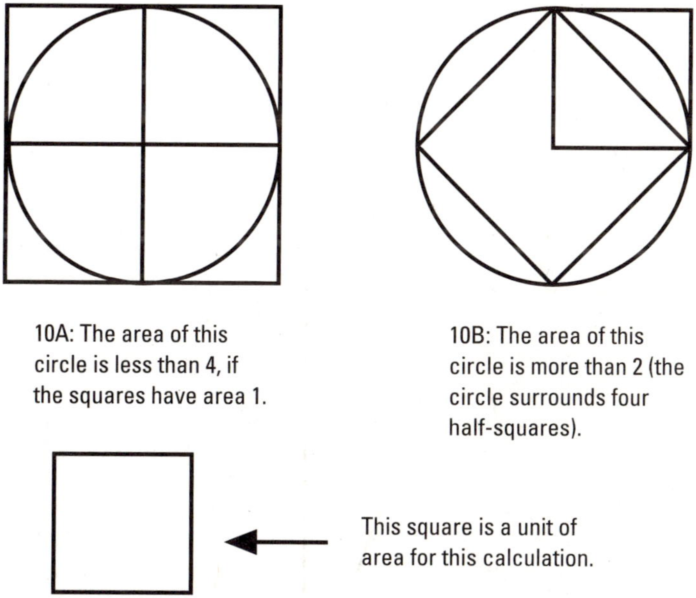

Figure 2-10: Estimating the area of a circle by using squares.

- If you draw a circle inside a square (10A), you can conclude that the area of a circle has to be smaller than four times the area of one of the little subsquares in the figure. If the area of a little subsquare is one, then the area of the circle is less than four.

- If you draw a square inside this same circle (10B), you conclude that the area of the circle has to be larger than the area of the four little sub-squares. The little sub-squares in this part of the figure have areas of ½ area unit, compared to the subsquares in 10A.
- In these units, the area of the circle is somewhere between four and two since those are the areas of the outside square set and the inside square set. In this diagram, the side of one of the squares in 10A corresponds to the radius of the circle (the radius goes from the middle of the circle to the edge). Following traditional Mesopotamian logic, looking at the numbers four and two, you would make an offering to the memory of Gilgamesh and say that the area of a circle is about three times the square of the radius. In modern terms, the rule would be

$$A = 3 \times r^2$$

In geometry, nothing to do with circles ever has to do with the number 3. It always has to do with the mystery number π, which is now known to be 3.141592653..., on and on with seemingly random digits.

Three is a nice, clean number, but even the ancients knew that they weren't getting away this easy. Using squares to try to cover a circle is a pretty sloppy procedure. Look at the diagram in Figure 2-11 for an improved covering, also known in ancient Mesopotamia and Egypt. By arranging triangles into hexagons (that nice, six-sided, turn-of-the-century bathroom tile shape is a hexagon) and taking area averages of the outside hexagon and the inside hexagon, the value for the area of the circle starts to look approximately like 3.15 times the radius of the circle squared. 3.15 is in fact a pretty good guess at the real value of π. It means that all these ancient calculations on circles were good to within 1 percent, which is probably better than the original accuracy of a knotted-rope measurement, anyway.

Both the Babylonians and the Egyptians had a distinct preference for expressing fractions only in the form

$$\text{fraction} = 1/\text{something}$$

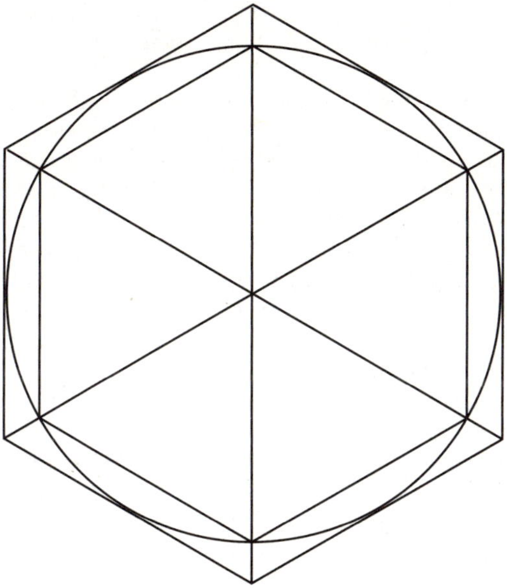

These hexagons cover the circle more exactly than squares do.

Figure 2-11: A better fit for a circle's area.

The fractional part of this number looked like about $\frac{1}{7}$, so the number $3 + \frac{1}{7}$ came to be widely accepted in the ancient world as a good working value for π (although nobody called it π, a Greek letter, until the Greeks took over the subject). This value was resoundingly good enough for two reasons:

- If you are measuring areas outdoors with knotted ropes, you probably only need to get within one percent of the correct area.
- Nobody in the ancient world had circular fields, so the whole exercise was a bit theoretical anyway.

And now, turn up the volume

So there are ancient formulas for areas of rectangles, triangles, and circles. One last useful quantity to estimate is volume, since temple tributes in the ancient world were often stated in terms of a certain volume of grain. In fact, your typical tax problem in the ancient world was that the local authorities kept a record of the area of land that you had under cultivation, and had worked out an estimate of the volume of grain that you were likely to produce from it. And also, of course, the officials estimated how much you owed them for their services in keeping order and regulating rainfall and river floods with prayer.

As you might expect from a down-to-earth bunch like the inhabitants of ancient alluvial valleys, the first rules for volume were probably worked out by using little models. After all, these people were up to their eyeballs in mud bricks, objects that lead directly to speculation about volume. If you think about it for a minute, you will realize that if you know how many bricks went into a solid brick temple platform, you automatically know the volume of the platform.

For two simple volume objects, the brick and the cylinder, the ancients realized that you just multiply an area times a height. Probably by direct experimentation (you can do the same thing yourself with a big box of sugar cubes), they apparently also worked out the volume of a cone (see Figure 2-12).

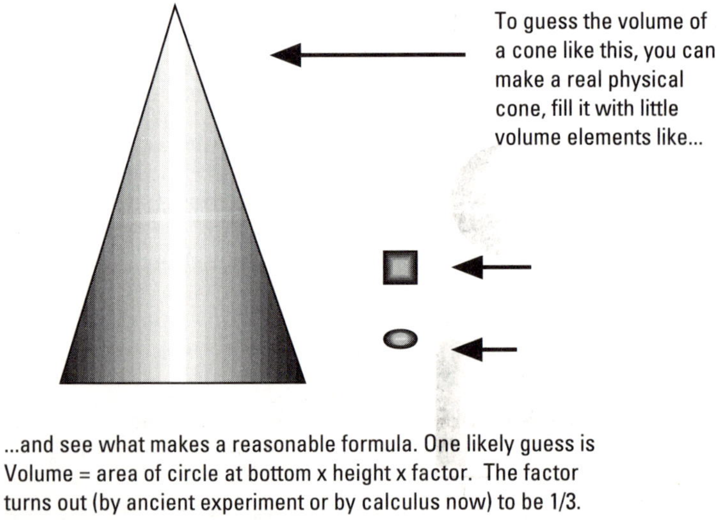

Figure 2-12: Volume formulas.

With the simple repertoire of area and volume formulas summarized in Figure 2-13 and a certain willingness to break eggs when making an omelet (you didn't want to get on the wrong side of an Egyptian tax collector, who was supposed to make up any shortfalls from his own account), it's possible to keep together a civilization that lasts many centuries. There was little philosophical content to the scribes' rule books, but they could tell you how many bricks you would need to beef up the city walls against the invaders from the desert.

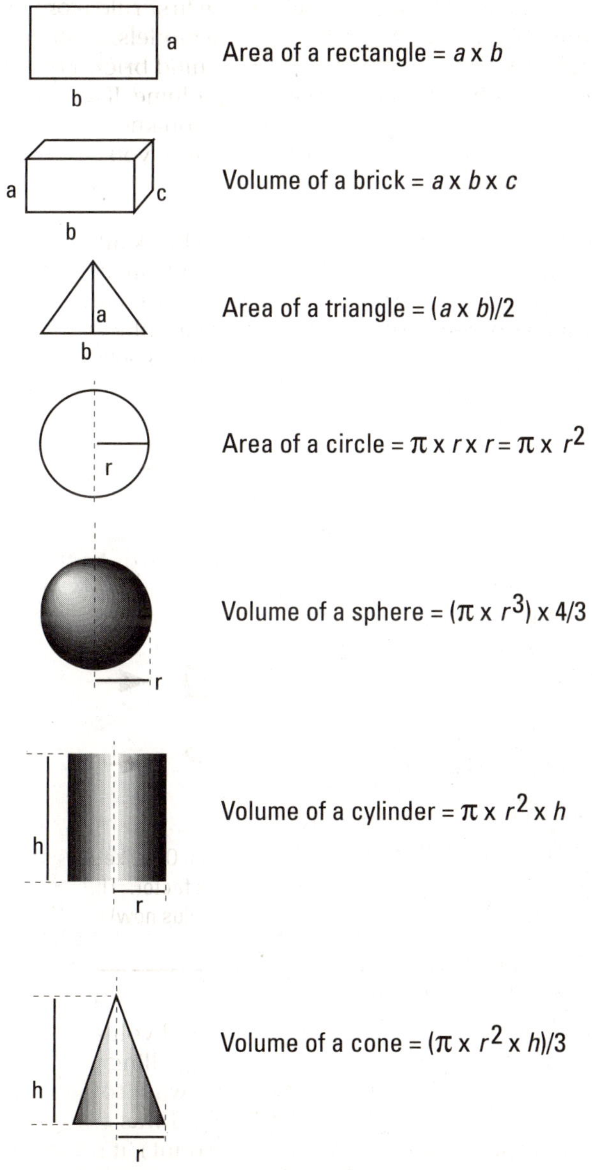

Figure 2-13: Simple formulas.

Philosophy Enters the Picture

Greek contact with the knowledge of the ancient world changed everything. You may notice that the first part of this chapter is probably easier to follow than Chapter 9 of your high school geometry book. That's because the Sumerians, Babylonians, and Egyptians just wanted to get their measurements right in a practical sense. By the time of Greek ascendancy in the eastern Mediterranean, the most useful rules of geometry were already 2,000 years old.

The Greeks were interested in the Why of things, in a close examination of fundamentals, and typically delighted in hair-splitting, exasperating argument. Culturally, they were what you call a Hard Act to Follow in several senses. It's hard to imagine anyone else in the ancient world arguing about Good or Virtue — elsewhere, Good simply meant having enough to eat for the first few thousand years of recorded civilization.

I don't propose in one chapter in a book called *Math Basics For Dummies* to review to any extent the idea of axiom-based proof, but I hope to suggest how Greek civilization transformed mathematics from a down-to-earth counting enterprise into the abstract activity it subsequently became. I have been encouraged to do this by a committee of high school math teachers, who figure that I can spare a few pages out of this book for general cultural literacy

Pythagoras

The Greek philosopher and mystic Pythagoras (who lived in southern Italy around 550 B.C.) is best known in geometry textbooks for the Pythagorean theorem (quoted exactly in the movie *The Wizard of Oz* by the Tin Man as he's awarded his diploma). Actually, a form of the theorem was known to earlier civilizations with their funky old knotted ropes (see Figure 2-14), but only for particular triangles.

It was known in the earliest civilizations that when the sides of a triangle fit the ratio 3, 4, 5, the triangle is a right triangle.

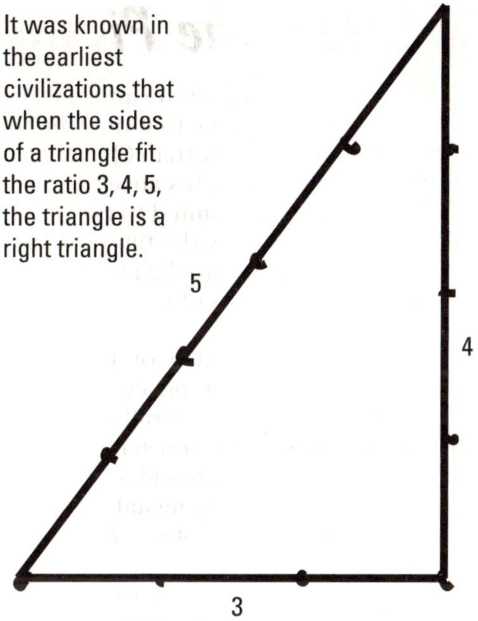

Figure 2-14: A 3-4-5 triangle in rope measurement.

There's reason to believe that the main Pythagorean result about right triangles (triangles in which one corner is a 90-degree angle, or right angle) was known for a handful of special right triangles. It's characteristic of Greek intellectual effort that Pythagoras (or his followers, or someone in the neighborhood) sought to generalize the old knotted-rope result to all right triangles. The theorem, with a diagram used in a standard proof, is shown in Figure 2-15.

There's a nifty proof of the same result from China (around 40 A.D.). Einstein also discovered a proof of his own when he was a teenager. I'm particularly impressed that one proof was also discovered by U.S. President James Garfield. I confidently expect that we will never again see anyone at that political level who can balance his or her own checkbook, much less come up with an original proof of a classic math result.

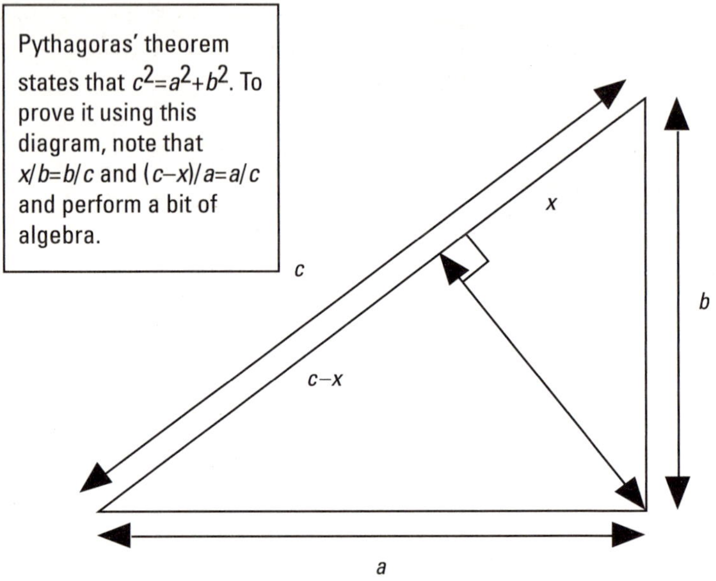

Figure 2-15: The Pythagorean theorem.

Zeno

Like Pythagoras, Zeno came from the Greek colonies in Italy (around 480 B.C.) and made it to Athens in time for the main events of the Golden Age when Athens was the leading city of the Greek world. He contributed some interesting paradoxes that succeeded in muddying the intellectual waters on questions of motion and time until the time of Newton. You can only get so far by thinking about things and denying common sense, although this originally Greek pastime is alive and well in many locations today. Find yourself a coffee house near a university if you don't believe me.

In one paradox, Zeno claimed that, basically, you couldn't get where you were going. First you would get halfway there, then you would get half of the rest of the way, then half of the rest of that, and so forth. This process of covering the next little bit of distance clearly would take forever (see Figure 2-16). In a related paradox, Zeno argued that the mighty Achilles could never overtake a tortoise with a head start, because in the time it takes Achilles to get to the tortoise, the tortoise has moved on, so there's another interval to close. Zeno's paradoxes were meant to suggest that you couldn't rely on the evidence of your senses. Even if you could catch up to a tortoise, Zeno could prove that you couldn't. So there.

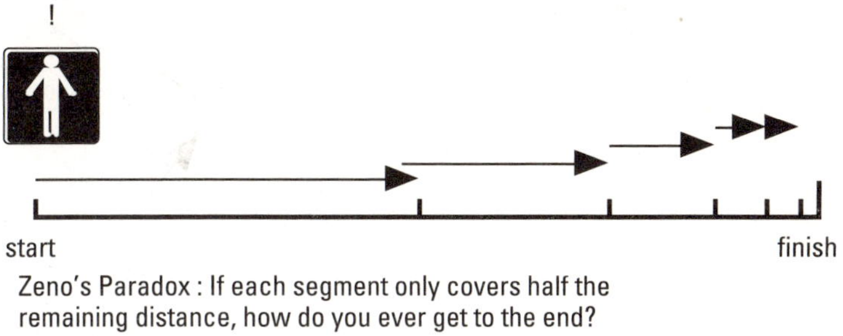

Figure 2-16: Zeno: Always getting halfway there.

The resolution of these issues is that it doesn't take an infinite amount of time for all these successively smaller gaps to close. Later in this chapter, in looking at Descartes' contribution to geometry, I'll revisit Zeno's paradox, mainly because I have always found it annoying that it was taken seriously for centuries. What was the matter with these people?!

Euclid

Euclid, who lived around 300 B.C. in a Greek world that had been transformed by Alexander the Great and wrote 13 books called the Elements, the first few books of which still constitute, in somewhat watered-down form, the first half of high school plane geometry.

In Euclidean geometry, you simply start with a few undefined geometric elements and a small set of axioms (see Figure 2-17) and proceed to derive all the geometric results of the previous few thousand years.

The emphasis of this effort is quite different from that of earlier geometry. For one thing, nothing is measured. Constructions are made using only a straight-edge (for drawing straight lines) and a compass (for drawing circles). You aren't allowed to make absolute measurements on a diagram — you might, for example, prove that two angles in a diagram are equal (see Figure 2-18), but you never actually measure the angles. Similarly, measuring areas or distances is "cheating" in this form of geometry — questions must be settled by argument rather than simple recourse to an experiment on paper or with materials.

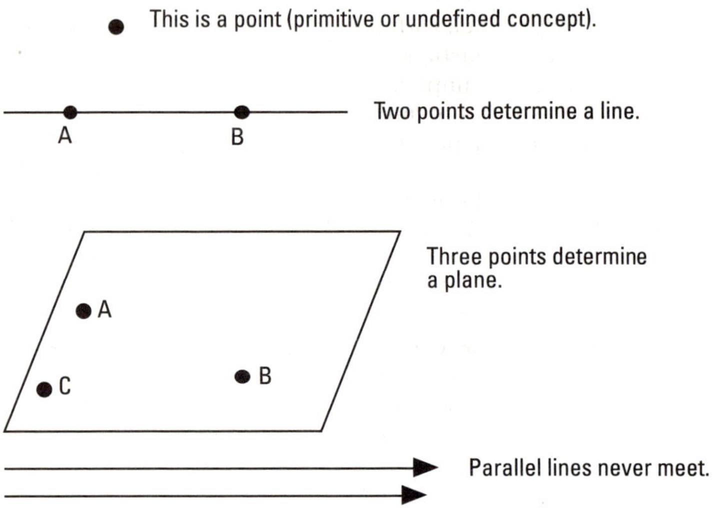

Figure 2-17: One page of Euclidean basics.

You can get the flavor of this enterprise, and something of its sociological motivation, by quickly examining this proposition:

The ratios of the lengths of corresponding sides of equiangular right-angled triangles are equal.

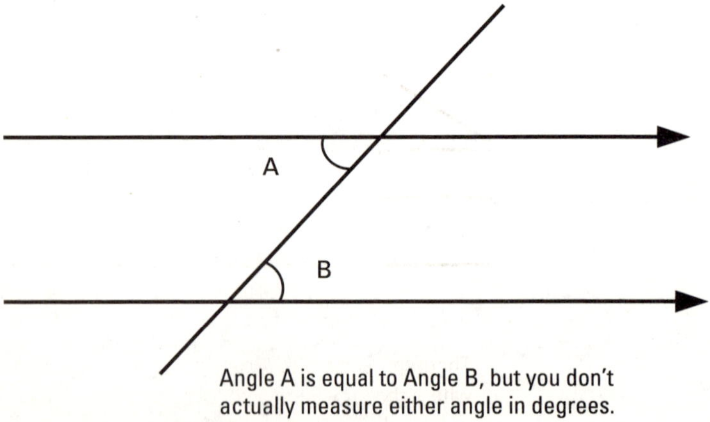

Figure 2-18: A classic Euclidean problem.

This statement has lots of practical consequences, as shown in Figure 2-19, but is sufficiently annoying to prove that it was known for centuries as the Pons Asinorum, in Latin the "Bridge of Donkeys," meaning that you're on the wrong side of the bridge (back with the donkeys) if you can't prove it. I believe that this particular bit of terminology calls into high contrast the rather unpleasant "proving you're smarter than the other guy" aspect of much of mathematics education, and a long-standing aspect it is, too.

In high school classes, it is seldom remarked that the ambitious 19th-century program to axiomatize, or make rigorous, all of mathematics, led to an interesting discovery. The discovery was that this can't be done. Greek mathematicians may have thought that every important result in geometry could be proven from a small set of axioms, but it has subsequently developed that this belief is wrong. This certainly doesn't mean that the idea of proof is outdated or that learning to prove theorems is a waste of time, but it does mean that the real world of mathematics has turned out to be a lot stranger than the Greeks imagined.

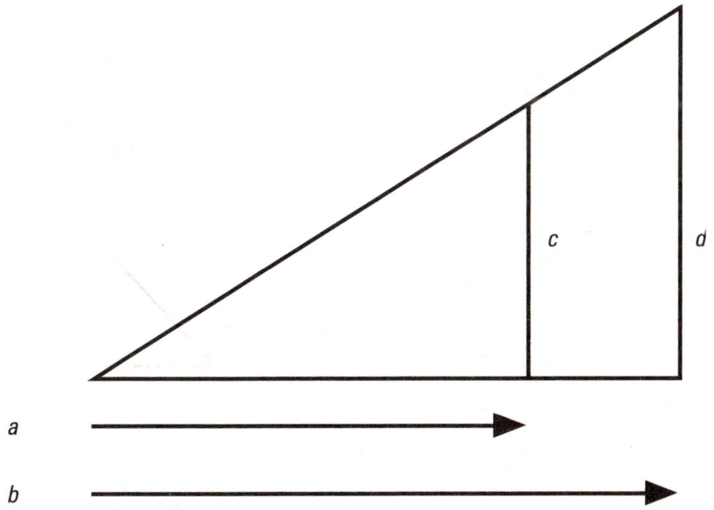

Theorem: Show that $c/a = d/b$, where the letters stand for lengths of the indicated line segments.

Figure 2-19: Using a theorem of Euclid.

The New World of Geometry

The first real innovation in geometry in close to 2,000 years occurred when the brilliant French philosopher René Descartes began to explore the idea of assigning numbers to points in a plane. Even if you aren't familiar with this idea, you use essentially the same principle every time you look up Whatzis St. on a city map and find it in D-6. In fact, a quick inspection of Figure 2-20 may convince you that Descartes' big idea was sitting there waiting to be discovered in Egypt.

When Descartes, loafing around in his bed all day in exile in Amsterdam (around 1630) developed the subject of analytical geometry, he really founded a whole family of subjects that form the basis of applied mathematics today. By analytical geometry, mathematicians really just mean geometry with numbers in it — the geometry of graphs.

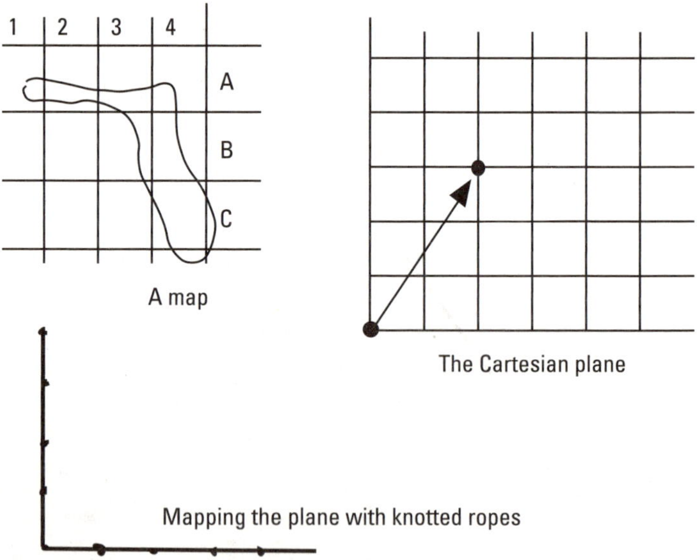

Figure 2-20: A map, a Cartesian plane, and two knotted ropes.

In the geometry of Euclid, there are essentially no numbers — the notion of quantity is expressed in terms of ratios. That's not how you design airplane wings. In Cartesian geometry, by contrast, numbers are everywhere, even to the extent that you can use numbers to define space with more dimensions than you can actually see. For this book, I just use examples from the plane, but everything that I say in this section applies to seven-dimensional space just as well.

Distance in a plane

Here are a few conventions in Descartes' geometry, called Cartesian geometry. The "up" direction in a Cartesian plane is usually called the *y*-direction, and sideways is the *x*-direction. Every point in the plane is identified by its *x*-position (called the *x* coordinate) and its *y*-position (called the *y* coordinate). The point (3,4), for example, means the point $x = 3$ and $y = 4$ — there it is in Figure 2-21 — you go sideways three units and then up four units to locate it.

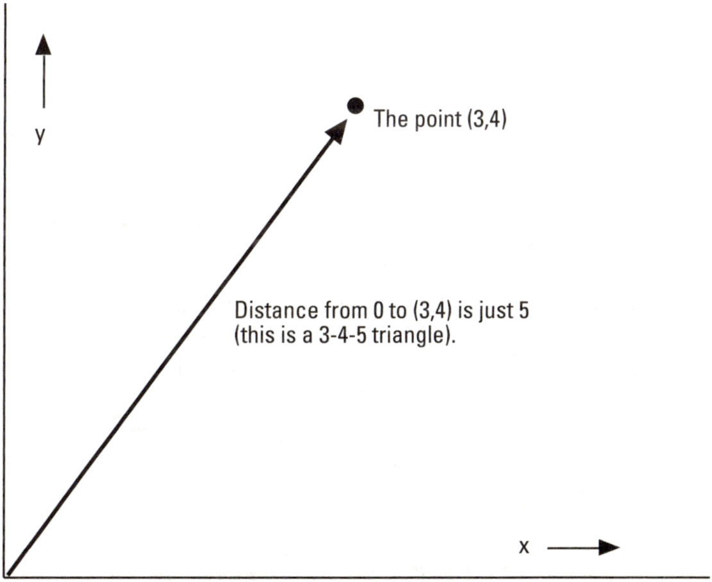

Figure 2-21: The point (3,4) and its distance from 0.

If you look at the figure again, you may see the outlines of your friend the Pythagorean theorem. In fact, you use this theorem to compute distance in the plane. Between every two points is a line that turns out to be the long side (hypotenuse) of a right triangle. Because it's easy to calculate the other two sides of the triangle, you can calculate the long side as well; the long side is just the distance between the two points (see Figure 2-22).

A line or two

Now, since I'm calling the first number in a pair the *x* coordinate and the second number the *y* coordinate, every time I write down some relation between *x* and *y*, I have defined something that turns up in the Cartesian plane as a piece of geometry. Consider two examples.

First, what is the meaning of this expression?

$$y = 2x$$

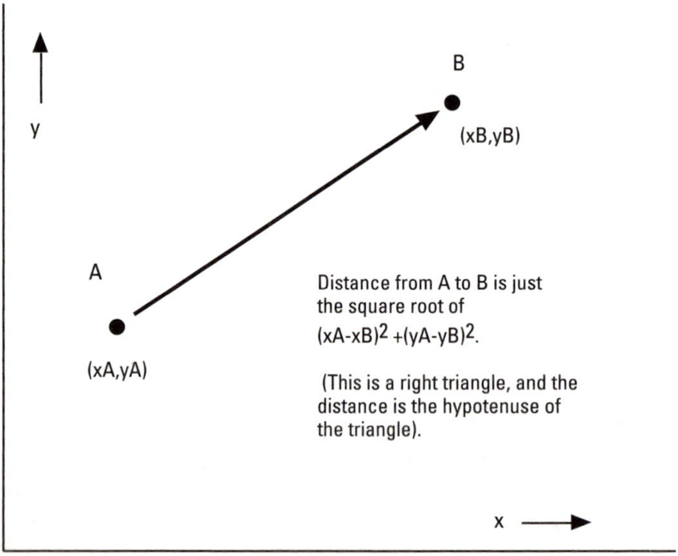

Figure 2-22: The distance between two points.

To find out the meaning, you pick a bunch of *x*-numbers, find the corresponding *y*-numbers, and then locate all these (*x*,*y*) pairs on the plane. This process should be pretty simple. Pick 1 for the first *x*. The expression says that the *y* that goes with this *x* is just 2*x*, which is 2. So one of the points is (1,2). Pick 3 for the next *x*, and the expression says that 6 is the corresponding *y*. That point is (3,6). In this way, you just make a short list of points and see where they fall on the plane. Check it out in Figure 2-23.

There. That's not so bad. You have now plotted a straight line in the Cartesian plane. Didn't hurt a bit. Emboldened by this success, try just one more modest advance.

Consider this expression:

$$y = x - 2$$

What happens here? For $x = 0$, the rule says that $y = -2$, giving you the point (0,–2). For $x = 8$, the rule says that $y = 6$, giving you the point (8,6). Collecting a few more points, you can make the plot in Figure 2-24. Note that the rule doesn't care if you put negative numbers like $x = -17.1$ into the expression — it's all the same from the point of view of plotting.

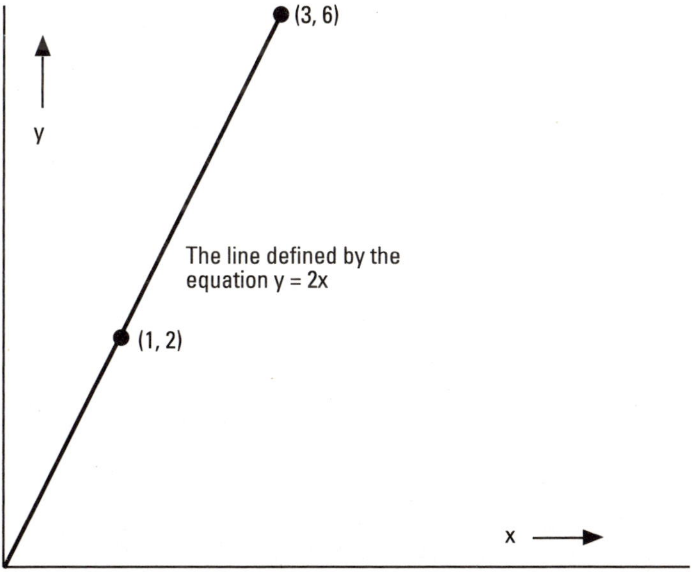

Figure 2-23: $y = 2x$ plotted in the plane.

It happens that every straight line in the plane is defined by an expression relating *y* and *x* this way (in a little bit, you'll see what happens when you have x^2 or something instead of plain *x*). All the constructions of Euclidean geometry involving straight lines can be discussed in terms of these *x*-*y* equations.

Not only does this make all sorts of proofs almost straightforward, but it allows a nice extension to curves as well, as you'll see in the following section.

A circular argument

In Figure 2-25, I have drawn a circle around the point (0,0). How would you describe this in *x-y* terms? Well, the definition of a circle is that every point is the same distance from the center. A few paragraphs back, I also gave a definition of distance.

In this case, the center is zero and the distance is always five units, so I'm going to try cooking up an expression this way:

$$\text{distance from } 0 = 5$$

Using the distance formula in Figure 2-23, you have

$$\text{area} = \text{length} \times \frac{(\text{short side} + \text{long side})}{2}$$

or

$$x^2 + y^2 = 25$$

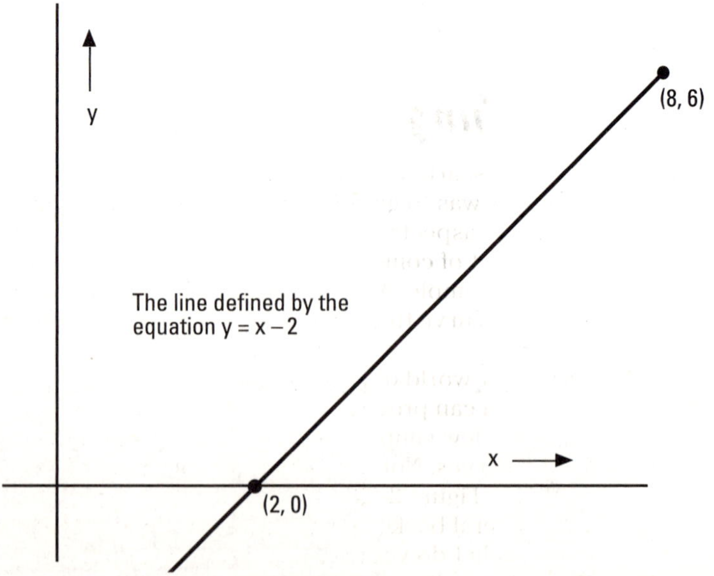

Figure 2-24: $y = x - 2$ defines another line.

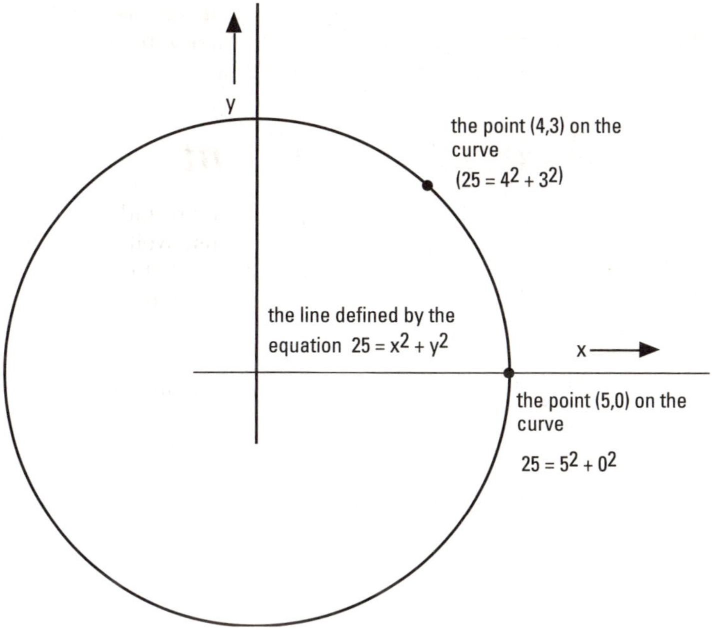

Figure 2-25: A circle in the Cartesian plane.

You can get out a calculator and show for yourself that points (x,y) that satisfy this expression all lie at exactly five units from zero in the plane.

Graphing

One of Descartes' great accomplishments in founding analytic geometry was to greatly reduce the strain on your own brain that some aspects of geometry induce. In the world of Euclid, the subject of conic sections was fraught with notable difficulties, for example. A conic section (ellipse, parabola, or hyperbola) is a curve that you get by slicing a cone with a plane.

In the new world of finding points from a rule and plotting them, you can prove all sorts of Euclidean theorems by performing a few simple algebraic operations on the expressions for the curves. Not much of this subject counts as everyday math, but Figure 2-26 shows what these curves look like for your general background. There's nothing in these plots that you couldn't do yourself just by making up the sets of points dictated by the expressions and then dumping those points onto the plane. Relatively inexpensive graphing calculators can do all this work for you, anyway.

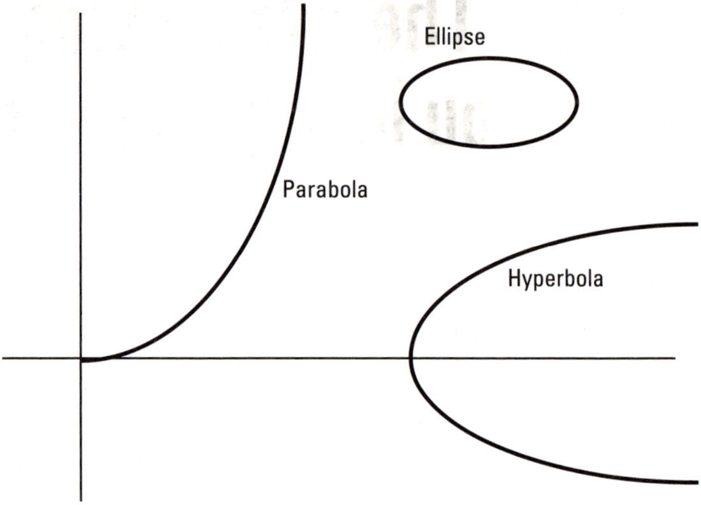

Figure 2-26: Three suspicious plots.

Chapter 3

What They Are Trying to Tell You in Trigonometry

In This Chapter

- Working with basic triangles
- Placing angles on a globe
- Relating sines, cosines, and waves
- Doing the wave

Trigonometry, often in the form of a course called Advanced Algebra, is one of those courses that typically distinguishes college-bound from non-college-bound students in high school. If you haven't taken it yet, or have forgotten nearly everything, I have a little good news for you: All the hard parts of trigonometry have been made fairly simple with the introduction of the scientific calculator.

This chapter should probably be titled "Cultural Literacy about Angles" because it discusses lots of topics but doesn't pose much problem-solving. For example, you wouldn't think that you'd find stuff about radio in a trigonometry chapter, but that's just one good example of the way all sorts of math topics are interrelated.

In the Land of Sines and Cosines

In Greek geometry, you spend lots of time proving that some angle is equal to another angle, but you never have an actual measurement, unless of course it's a right angle.

Egyptian geometry places a big emphasis on having a number for the angle under discussion. In terms of how many blocks of limestone you need, it makes a big difference whether a pyramid is going to be slightly flatter or slightly steeper. (*Slightly* is the right word here, too, because all pyramids use nearly the same angles.) The Egyptian measurement of angle is the *seked* (see Figure 3-1), and the sides of pyramids have angles between 5¼ and 5½ *seked.*

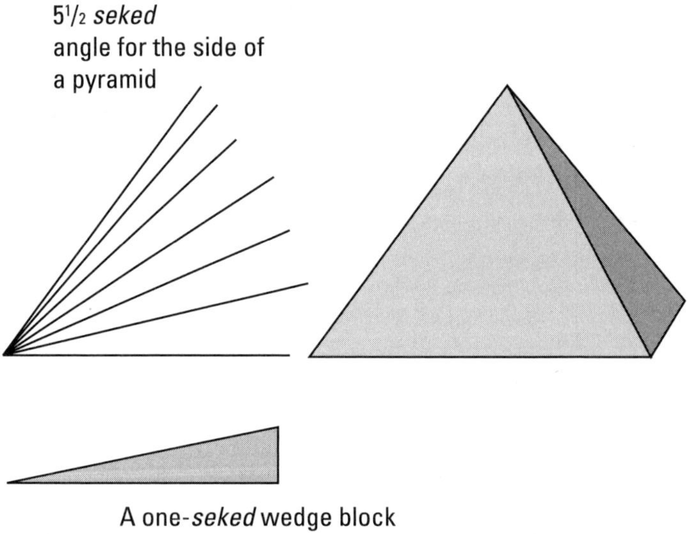

Figure 3-1: The great pyramid, a 5½-*seked* creation.

The practical Egyptians probably liked the *seked* because it's a good angle for a wedge or a ramp. In this civilization, in contrast, we divide the angles of a whole circle into 360 degrees. We inherited this number of degrees from the Babylonians, who were fond of multiples of 12 and 20. Like the early astronomers of Mexico, the Babylonians felt that the year should have had 360 days because 360 is such a cool number. Practically every early civilization had the year organized into 360 days that were okay and 5 bad-luck days stashed at the end. With 360 degrees in a circle, you have the tableau of appealing and popular angles shown in Figure 3-2.

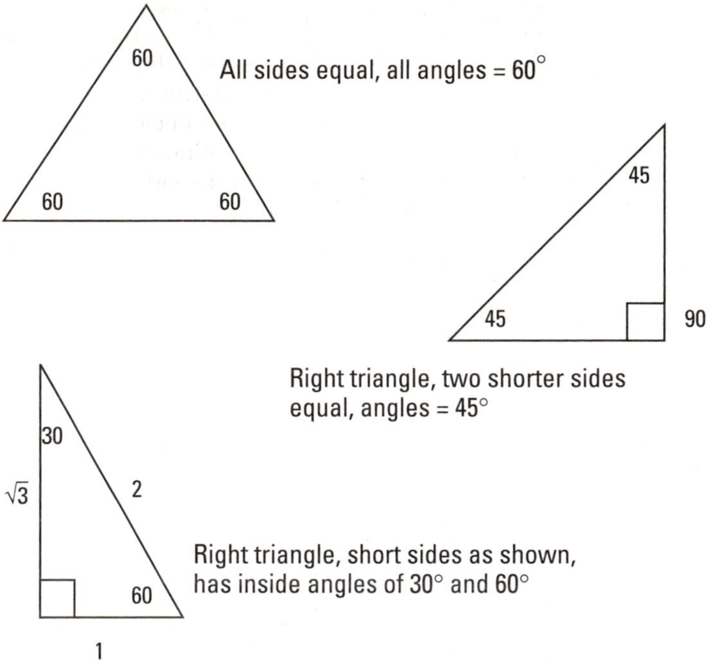

Figure 3-2: A few angles you should know.

Triangles in a circle

If you have a little one-*seked* right triangle, you can guess that a very big one-*seked* right triangle will be in proportion to it. (See Figure 3-3.) In 2000 B.C., the Egyptians knew that they could use this trick to estimate distances they couldn't actually measure, and that's just what they did (see Figure 3-4).

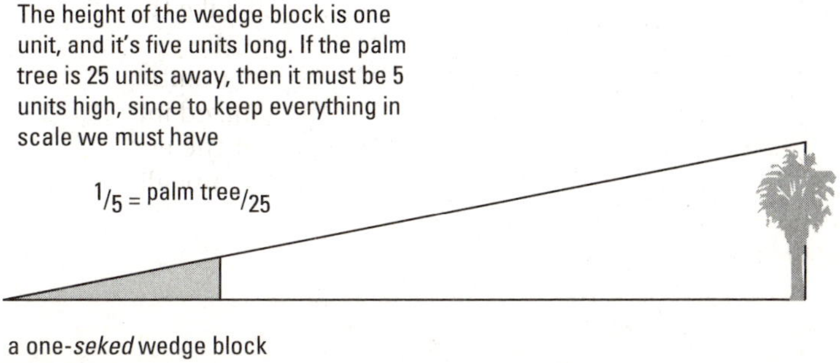

Figure 3-3: The Egyptian triangle gimmick.

The question facing you is this: Can you use this proportional-triangles trick for any angle? The answer provided by a series of mathematicians in the 17th century is: You bet you can!

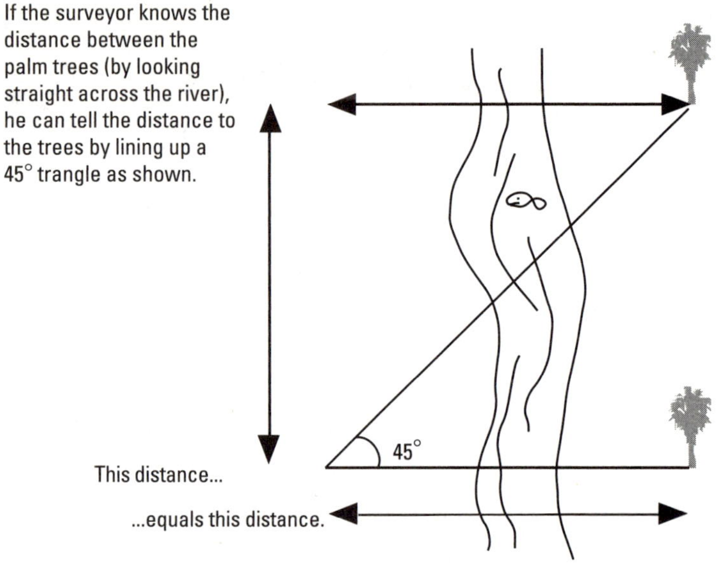

Figure 3-4: The triangle gimmick in surveying practice.

Here's how the process was generalized. If you draw right triangles inside a circle as in Figure 3-5, you make a whole series of triangles for which you know two things:

- The radius of the circle is fixed at one.
- s^2 and c^2 together have to add up to one, because of Pythagoras' theorem about right triangles.

Here I'm using my own names for the sides of the triangles — straight up is *s* and crossways is *c*. The real names for these triangle sides are sine (*s*) and cosine (*c*), abbreviated as *sin* and *cos*. Out of deference to the Greeks, the angle in this situation is usually called theta (θ), but I'm going to call the angle *a* because I am a plain-speaking person from the Midwest.

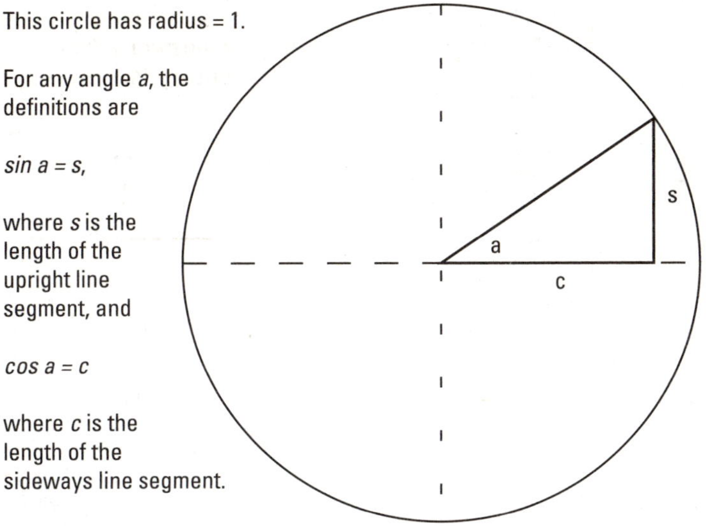

Figure 3-5: The triangles in a circle define sines and cosines.

Sine, cosine, and surveying

The number for sine depends on the angle a. Table 3-1 shows a little table of sines. Because the sine depends on the size of the angle, you say that sine is a function of angle a and write an expression like this:

$$0.5 = \sin(a) \text{ for } a = 30 \text{ degrees}$$

You change a, and $\sin(a)$ changes. That's the meaning of a function. If you know a, you can look up $\sin(a)$.

Table 3-1	**A Little Sine Table**
sin 10°	0.17
sin 30°	0.50
sin 45°	0.71
sin 60°	0.87
sin 90°	1.00
sin 180°	0.00

Now, how does any of this amount to an improvement on the old Egyptian Triangle Trick for surveying? Here's how: In trigonometry, if you have the length of just one side of a right triangle and the length of a side of a live angle (any angle but the 90-degree angle), you can calculate the other sides. In the first version of Trigonometry Against the Pyramids, you get a diagram like the one in Figure 3-6.

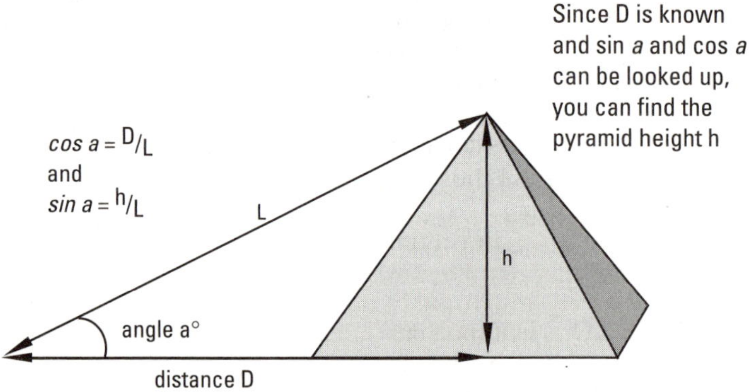

Figure 3-6: The great pyramid, now a trig problem.

But trigonometry has even fancier tricks than that. Any triangle can be decomposed into two right triangles, which means that it's possible to cook up formulas to cover all sorts of examples in surveying (see Figure 3-7), even if it's inconvenient to formulate the problem in terms of right triangles. And that's just what you see surveyors doing — measuring distances and angles to make maps.

The sides of this triangle have lengths A, B, and C. If you know two of them and can measure the angle *a*, then you can find the length of the third side from the rule

$A^2 = B^2 + C^2 - 2BC\cos a$

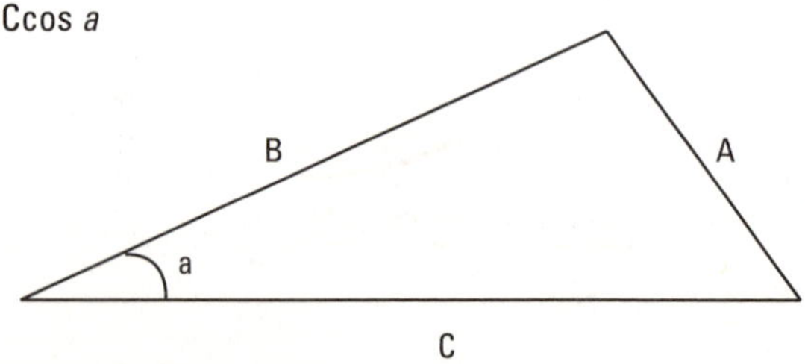

Figure 3-7: Trickier problems, trickier trig.

Angles on the Earth

The previous section talks about plane trigonometry, measurements taken on a flat surface, or at least a surface that's flat enough that curvature can be ignored. But angles can just as easily be defined on a sphere. As this is being written, my approximate location is 123 degrees west and 39 degrees north. We all live on the surface of a sphere, and the radius is essentially constant — even Mount Everest, at 5 miles high, amounts to a tiny part-per-thousand surface blemish on this smooth globe with a radius of 4,000 miles. That means that our location in three-dimensional space can be closely specified by just two angles, as shown in Figure 3-8. So how did people find these angles out on the featureless sea?

By late medieval times, mariners had figured out that they could determine how far north or south they were on the planet by sighting the sun with an instrument that measured the angle of the sun above the horizon at noon. (Despite what you may have been told, by 1350 or so, navigators in Europe and the Arab world not only knew the earth was round but they also had a pretty good idea of its diameter.) How does the sun's angle relate to north and south?

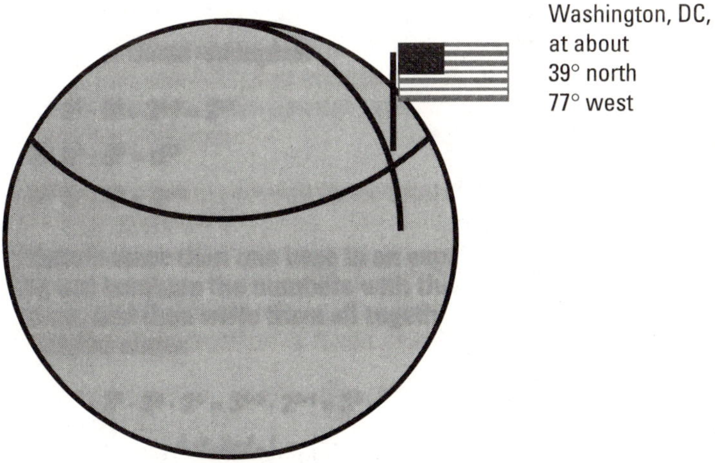

Figure 3-8: Home, with angles.

Up, down . . .

Here's how. Picture yourself on a sunny day in March, standing around doing nothing near the North Pole (remember, there aren't even penguins to play with). At midday, you sight the sun and you're looking practically straight sideways. Now picture yourself lounging around Quito, Ecuador (why do you think they call it Ecuador?), with a Pisco Sour in your hand, sighting the sun at noon from the front of a fake-Incan gift store. That's right, the sun is directly overhead.

So anywhere in between, from the equator to the North Pole, you would find an angle somewhere between straight up (90 degrees) and straight sideways (0 degrees). That gives you a latitude (the number of degrees north or south of the equator). The instrument used for measuring the angle looks just like a protractor from geometry class, and lovely brass examples taken from real ships can be found in museums everywhere.

FUN FACT

As a curiosity, you might ask what one angle is worth in terms of distance. The world has a diameter of about 8,000 miles, so according to the geometry rules in Chapter 2, you just have a circumference of π times the diameter, which is $3.14 \times 8{,}000 = 25{,}120$ miles. (The circumference of the world is the distance around the world at the middle.) Since there are 360 degrees in a circle around the world, one degree of latitude is worth about 70 miles (you divide 25,120 by 360).

All around

Finding the north/south location isn't the difficult problem. The problem is measuring the sideways direction, called longitude. You see, the planet is spinning, a great inconvenience in making some directional sightings. Plenty of stars provide reference points for you, but you're spinning around in the middle of them. If you had a star map and a way to tell time, you could perform your sightings at the same point every night.

Alternatively, if you had a time reference for one fixed point on the earth's surface and also your own clock, you could compare the time of sunrise in your location with the time of sunrise at the fixed point (you look it up in a table provided by navigation authorities). This pretty much tells you what your time zone is.

All these time zones are referenced to Greenwich, England, and longitude zero is assigned to a line running through the Royal Naval Observatory at Greenwich. (Remember, latitude zero is the equator.) Not coincidentally, the British Admiralty commissioned the development of clocks that could hold time accurately on board a ship ploughing through 10,000 miles of roaring Pacific storms. These clocks (called chronometers) made it possible to provide great accuracy in the angle of longitude, so little specks in the vast ocean, such as Niue or Norfolk Island (look 'em up!), could be placed definitively on a navigational chart.

Man-made stars

After centuries of cleverness in navigation, the whole business has now been cleaned up electronically for the benefit of absolute dunces. One of the main motivations of the first satellite builders was to launch a satellite-based navigation system that could give accurate locations anywhere, at any time.

The first systems to interpret the satellite navigational signals were the size of a refrigerator and cost hundreds of thousands of dollars — the appropriate vehicle for using these systems was an aircraft carrier. Now, years later, a calculator-sized navigational system that automatically reads out latitude and longitude with enough accuracy to place you within 20 feet or so costs about $300. I have a friend who bought one for his speedboat, used only for water-skiing on Lake Sonoma. You can't get lost on this lake — you can see the dock from any point, for example. But it was so inexpensive (from a hardware store, too — not even a proper marine supply store) that he couldn't resist. You can buy one and track your trip to Grandma's for Thanksgiving in angular coordinates on the planet from start to finish.

Sines and Waves

And when you get back from Grandma's, trigonometry enters your home in a few interesting ways — literally, in the form of sound waves, radio waves, and all sorts of other phenomena.

Simple waves

Figure 3-9 is a plot of sin(a) as a runs through all the angles from 0 to 360 degrees. At 360 degrees, of course, the angle is right back where it started, so the next part of the plot looks just the same (see Figure 3-10). What you're looking at in this plot is the simplest kind of wave.

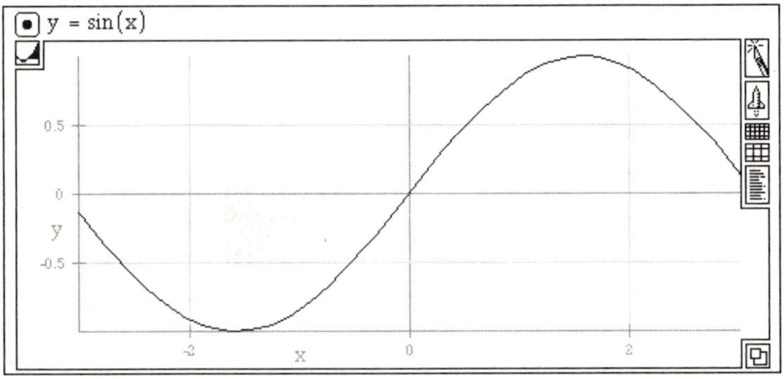

Figure 3-9: One cycle of a sine wave.

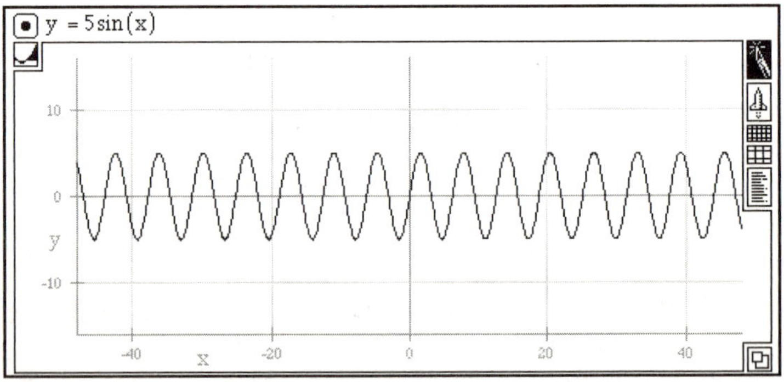

Figure 3-10: A wave extending for many cycles.

Pingggggg: Sound waves

If you strike a tuning fork, it sends a sound pressure pattern out into the room that matches this sort of wave. A sine wave, as it's called, corresponds to a pure single tone. If you actually measure the tone with a special microphone, the height of the peaks shows how loud the tone is, and the spacing between the peaks gives the wavelength. Wide spacing means a low note; narrow spacing means a high-pitched tone (see Figure 3-11). In cycles per second (the measurement of frequency now usually called *hertz,* abbreviated Hz), you can hear tones down to about 30 Hz or so and up to perhaps 12,000 Hz. Higher tones, for example, 20,000 Hz, are in the range of dog whistles — you can't hear tones that high.

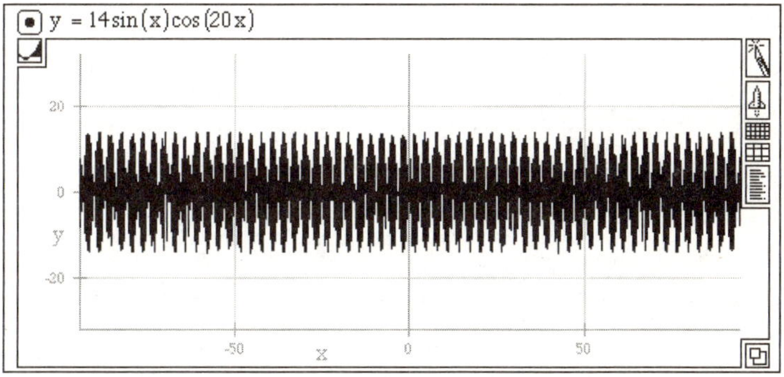

Figure 3-11: Mathematical music.

So you can have waves of different heights (heights are called *amplitudes* in wavespeak) and different frequencies. The feature of these waves that gives sound its variety is the rule that waves can simply be added to each other. Figure 3-12 shows a whole collection of waves added to each other. Any sound can be represented as a set of waves added to each other to produce a waveform. Complex sounds like speech have very complicated waveforms — that's one reason why computer recognition of speech is such a difficult endeavor. The sounds of real musical instruments, too, are never pure tones but complicated mixtures. That's why it took years of research to make synthesized sounds that sound reasonably close to the sounds of a real piano.

It also happens that cosines have their own waveforms, which look exactly like sine waves except that their zero points are shifted 90 degrees. You can add these to sine waves, and for that matter multiply sine and cosine waves, giving fantastically complex and interesting wave shapes. And every one of these shapes corresponds to a sound that you can actually hear. A music synthesizer just adds up waves electronically and then shapes the amplitude of the resulting waveform.

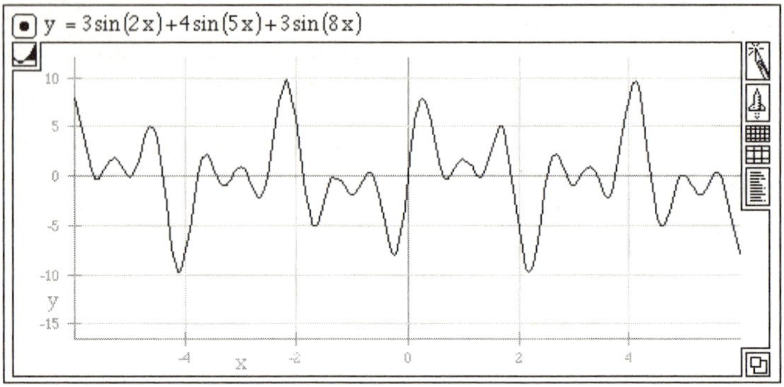

Figure 3-12: A wave library.

At the speed of light: electromagnetic sine waves

Every form of radiation, from sunlight to x-rays to AM radio, appears as electromagnetic sine waves that are comparable to sound waves except that

- An electric field is waving instead of air pressure, and
- The frequencies in Hz can range much higher than sound waves.

Just for another example of the tricks you can do with sine waves, I'll consider the case of AM radio. Here's the story: If you speak into a microphone, the sound waves are converted by the microphone into electromagnetic waves traveling down a wire. If you are singing a pure 500 Hz tone (you can't actually get it that pure), the microphone produces a pure 500 Hz electromagnetic wave in the wire.

The challenge of AM radio is to find a way to send this wave out on an antenna in a way that can be received miles away. It's done simply by multiplying the wave from the microphone with a much-higher-frequency carrier wave.

The basic idea is illustrated in Figure 3-13. For a station operating at 1,200 kiloHz (that's 1,200,000 Hz), the result is a wave that looks like a plain 1,200,000 Hz sine wave with an amplitude modified to display a 500 Hz ripple. The AM radio receiver can be tuned to subtract out the sine wave carrier, leaving an electrical 500 Hz signal to be passed to a speaker, where it's converted back to sound at 500 Hz.

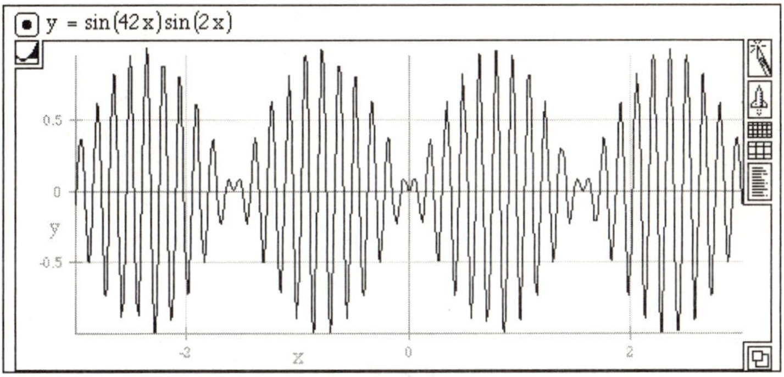

Figure 3-13: The grand plan of AM radio.

That's where AM radio gets its name; the AM stands for Amplitude Modulation. It's also the reason that AM radio isn't a first choice for music accuracy. Because there are stations spaced every 10,000 Hz or so along the dial, each station has to clip its music signals in frequency to prevent overlapping neighboring stations. There's plenty of signal space (called bandwidth) for talk, which doesn't require a big range of frequencies, but AM isn't going to do an outstanding job at violin concertos.

Anything you can pick up on any kind of antenna was originally assembled back at the station as a collection of sine waves added or multiplied together electronically, just like the sine waves in the little diagrams in this chapter. Half the tricks in modern electronics consist of ways to make the waves combine exactly without introducing distortion. And then, at the receiving end, you own equipment that can disassemble the waveforms back to a sound signal or a TV scan line or some other sort of message. The whole world of communications, including the satellite navigation signals previously mentioned, is based on nothing more than combining the humble circular functions of trigonometry.

Chapter 4

Percentages

In This Chapter

- Percentages and fractions
- Mark-ups and mark-downs
- Useful percent tricks
- Percent idiocy in the news

Percentages, I'm afraid, are always around. They make a convenient way to concoct some fake drama from ho-hum statistics, and they make it possible to make deals at a sale seem better than they are. Traditionally, interest rates are expressed as percentages, introducing one more little bit of confusion into calculations with financial calculators. The numbers themselves can tell stories quite eloquently, but percentages sometimes rear their ugly heads when there's incentive to make the story bigger, smaller, or fuzzier.

Everyone could have agreed, long ago, that all references to numbers would use just the numbers themselves. Here's how that would work:

- You go into a clothing store and they're having a sweater sale. The sign would say, "$39.99 sweaters now cost $19.99. Don't worry about what it says on the tag — our cash registers are programmed to get it right."
- At the end of the year, the police chief of Nothingville reports that there were three home robberies in the last year instead of two. The Nothingville Times-Dispatch could then headline, "Three Home Break-Ins Last Year."

- You read a psychology report in the newspaper. It states that in a sample of 20 people tested, 10 coffee drinkers and 10 non-coffee-drinkers, 6 of the coffee-drinkers chewed their pencil erasers while only 5 of the non-coffee-drinkers did.

These little examples contain all the information you need in the numbers of their respective stories. Now read the equivalent statements in percentages and decide for yourself whether the statements become any clearer:

- "Sweaters Now 50% Off — Discount Taken at the Register"
- "Home Break-Ins Up 50% in Nothingville"
- "Coffee Held Responsible for 17% Increase in Eraser-Chewing"

Percentages and Fractions

A percentage is just a fraction. But it's a fraction expressed as a fractional part of the number 100 — it's from a Latin term per centum that means "per 100." Explanations of percents result in some major weirdness in math books. Please let me explain what I mean by weird in this context.

Practically every math book on the market offers as a tip or trick that you can find 25 percent of some quantity just by dividing that quantity by four. For example, 25 percent of the number 32 is just 8 because $^{32}/_{4}$ is 8.

This isn't a tip.

This isn't a trick.

Twenty-five percent is one-fourth.

Twenty-five percent is just a slightly more confusing way of expressing the number one-fourth. The ancients, who had no calculators, had a distinct preference for expressing fractions in the form (1/number) because they used fractions as directions for dividing actual physical objects (grain, bricks, or buckets of olives, for example). If you have a pile of bricks, it's pretty easy to make $^{1}/_{3}$ of the pile by arranging the bricks in three lines. If you want to find 33 percent of a pile of bricks, you have to convert the percentage to a fraction anyway.

Table 4-1 shows common fractions and their equivalent percentages. Realistically, in a world of $2 calculators, you need to know only a few of these numbers. You should know that 50 percent is the same thing as one-half — hey, you probably do know that 50 percent is one-half. But for the rest of the percentage problems in this chapter, I recommend that you do whatever it takes to get the right answer (I'll tell you how), and more often than not, that means converting each question to a foolproof calculator problem.

Table 4-1 Values Expressed as Fractions and as Percentages

Fraction	*Percent*	*Decimal Fraction*
$\frac{1}{100}$	1%	0.01
$\frac{1}{8}$	12.5%	0.125
$\frac{1}{4}$	25%	0.25
$\frac{1}{3}$	33.3%	0.333
$\frac{1}{2}$	50%	0.50
$\frac{2}{3}$	66.7%	0.667
$\frac{3}{4}$	75%	0.75
$\frac{5}{6}$	83.3%	0.833

If you have to remember one fast rule (again, it's a definition, not a trick), remember this:

To convert a percentage to a decimal fraction, move the decimal point two places to the left. Thus

$$18\% = 0.18$$

$$37.5\% = 0.375$$

$$100\% = 1.00$$

Shopping

Enough of this abstract palaver. This book hopes to keep your attention by focusing on money. Let's go to the store and decode some common situations.

Everything 20 percent off

When everything in a store is marked down by 20 percent, it means two things, both of which are the same, mathematically speaking:

- Everything has been marked down by $\frac{1}{5}$ of the original price.
- Everything now costs $\frac{4}{5}$ as much as it did before.

Now try some arithmetic. You go to an auto parts store, where a small socket-wrench set cost $17.99 before the sale. Now it's 20 percent off. Twenty percent is not just $\frac{1}{5}$; by definition, it's 0.20, expressed using a decimal point.

First, multiply 17.99 by 0.20.

$$17.99 \times 0.20 = 3.598$$

What does 3.598 mean? It means $3.60, because stores have no way to charge you $\frac{8}{10}$ of a cent. So the new price is

$$17.99 - 3.60 = 14.39$$

Now, just to illustrate the second statement, note that all of something is 100 percent. If you take away 20 percent, what's left is 80 percent. This is the same as saying that $\frac{4}{5}$ is left when you take away $\frac{1}{5}$ (see Table 4-2). By using this convenient one-step method, you get a new price of

$$17.99 \times 0.80 = 14.392$$

Again, there aren't fractions of cents, at least in stores, so the price is $14.39.

It's kind of odd, really, that it never became the custom to say, "Everything in the store is now 80 percent of the old price." I guess the idea is that the sign should bring to mind the lump of money you keep, not the lump you pay.

Table 4-2 Fractions and Percentages in the Store

Original Price	*Discount %*	*Discount in $*	*Final Price*
$17.99	30%	$5.40	$12.59
$24.99	35%	$8.75	$16.24
$100.00	20%	$20.00	$80.00
$219.00	15%	$32.85	$186.15
$499.00	18%	$89.82	$409.18
$7.99	25%	$2.00	$5.99
$139.99	50%	$70.00	$70.00
$59.99	20%	$12.00	$47.99
$199.99	25%	$50.00	$149.99

Additional 25 percent discount taken at the register

EXAMPLE

Suppose you're in a department store looking at a jacket that started out at $79.99. The little computer-printed tag now has a line drawn through the $79.99 in red ball-point ink, and the price on the tag says $39.99. That means that the jacket has already been discounted by 50 percent.

Can you see why? Follow these steps:

1. **Round the original price to the nearest $10.**

   Think of $79.99 as $80, since, after all, that's about what it is.

2. **Round the sale price to the nearest $10.**

   $39.99, similarly, is close enough to $40.

3. **Compare the original price to the sale price.**

   The price went from $80 down to $40. In other words, it got cut in half.

   The fraction ½ and the percentage 50 percent are the same thing. Cutting the price in half means a discount of 50 percent.

But now you get an additional 25 percent off the marked price when you actually buy the jacket. What does this amount to? Figure it out in one of these two ways:

- One way to look at it is that you get a discount of 0.25 times the new price of $39.99. You can do this calculation pretty easily in your head if you recognize that $39.99 is $40 and 25 percent off means ¼ off. One-fourth of $40 is just $10, so the new price is going to be $30 (although in store arithmetic, that undoubtedly means that the new discounted price is $29.99).
- The other way to look at it is that 25 percent off means that 75 percent of the price is still there.

$$\$39.99 \times 0.75 = 29.99$$

You get the exact same thing, as you would expect.

Mark-Up

Stores, of course, buy all their merchandise at a discount from the price you see. In grocery stores, this mark-up, as it's called, is relatively modest. In jewelry stores, it's relatively outrageous, sometimes as high as 300 percent.

What's the meaning of a 300 percent mark-up? The store buys a ring at a price of $80. A 100 percent mark-up would be $80 added to the original $80, giving the ring a shelf price of $160. A 200 percent mark-up would be twice as much, or 2 times $80 added to the original $80, for a price of $240. And a 300 percent mark- up means a final price of $320.

Some of this is a bit confusing, which is often the case with percentages. In this case, it is actually simpler to say that merchandise gets marked up by a factor of four instead of referring to the same operation as 300 percent mark-up. A formula for this mark-up would be

$$\text{final price} = [1 + (\text{percent mark-up}/100)] \times \text{original price}$$

Plugging in the numbers for this jewelry example, you get

$$\begin{aligned} \text{final price} &= [1 + (300/100)] \times 80 \\ &= (1 + 3) \times 80 \\ &= 4 \times 80 = 320 \end{aligned}$$

Of course, this is one of the reasons that some jewelry stores post signs saying, "Pre-Christmas Sale! Pearls and Diamonds 50 Percent Off!" If the store marks up the merchandise by a factor of four when it first comes in the door, it can mark it back down by a factor of two and still make money.

Calculators and Percent Keys

Amazingly, there isn't much consistency in the way you use the % key on different calculators. Calculators from Texas Instruments, for example, use one style. If you want to find a 25 percent discount on $65, you press this sequence:

65

–

25

%

=

TIP

and you get the answer 48.75. I like this style because it follows the way you would state the problem out loud. As an unsolicited plug, I think that no home should be without a Texas Instruments BA-35, one of the least expensive financial calculators.

Alternatively, most Casio calculators do things a little differently. On a Casio, you do the same problem by pressing these keys:

65

×

25

%

–

This order seems a little strange to me, but it works. Of course, the idea that you would actually have to read the instruction booklet that came with your calculator is simply terrifying, but sometimes you just have to bite the bullet.

Ups and Downs

One odd feature of doing calculations in terms of percentages is that a change described as "down 25 percent" followed by a change of "up 25 percent" doesn't get you back to the original number.

Suppose there were 150 fatal traffic accidents in your county last year. (I hope this is because it's a big county and not because it's particularly dangerous.) Now, what's the number of traffic fatalities if the number goes down 20 percent?

$$150$$
$$-\ (20\%\text{ of }150\text{, which is }0.20 \times 150 = 30)$$
$$= 120$$

So a 20 percent drop in traffic fatalities gets you to 120 from 150.

What happens if, in the following year, the number of fatalities goes up by 20 percent? Remember that now you're starting with the figure 120 as the basis of the comparison. You get

$$120$$
$$+\ (20\%\text{ of }120\text{, which is }0.20 \times 120 = 24)$$
$$= 144$$

Ups and Downs in the News

This business about mark-up is closely related to one of the main uses of percentages, the daily announcement of percentage changes in some quantity or other.

Here's an example of how the calculation works. Last year, 15 cars were reported stolen in Burgertown, a city of 100,000 people. In the preceding year, 12 cars were reported stolen. There are two ways to calculate the percentage change in car thefts:

- Using a difference

  The net change in car thefts is 15 – 12 = 3, and the base for the change is a first-year figure of 12. So the change is

  $$\text{change} = 3 \div 12 = 0.25 = 25\%$$

- Straight calculator work

  Divide the number for the new year by the number for the preceding year. That's 15 divided by 12, or

  $$15 \div 12 = 1.25$$

  Last year was 100 percent, equivalent to 1. To compare the two years, subtract last year (1.25 – 1 = 0.25) and keep the 0.25. That's 25 percent. You can look at the number 1.25 and say that this year's thefts are 125 percent of last year, or you can say that this year's thefts are up 25 percent from last year.

The problem with this style of reporting as compared to having the original information is that you become accustomed to thinking of changes as significant, whereas the actual numbers are insignificant. In this case, a change from 15 to 12 reported car thefts in a town of this size really doesn't represent a real trend — that size of fluctuation would be expected. Also, this report doesn't point out the main fact, which is that car theft is not a big problem for this community.

All numbers in reported statistics fluctuate by a certain amount. One good estimate of the natural amount of variation from one sampling period to the next is

$$\text{variation} = \sqrt{\text{quantity}}$$

If an area reports 10 car thefts per year, it would be more or less expected that this number might naturally bounce up or down by as much as 3 cars from one year to the next. If your area records about 20 embezzlements per year, you can expect the number not to be exactly 20 every year but 20 plus or minus 4 or 5.

In other words, 18 embezzlements one year and 21 the next doesn't mean that an accounting crime wave has started. Rather, it just means that the number will change from year to year, sometimes increasing and sometimes decreasing, by a predictable amount. If, in this case, you get increases of more than five cases two years in a row, then it's time to see whether something strange is happening.

Part of the problem is that they're going to print that newspaper every day, whether there's really anything to report or not. That's why sports, weather, and the stock market are such prized news segments — they're guaranteed to provide some sort of fluctuating numbers no matter what.

Chapter 5
Statistics in the News

In This Chapter

- Distinguishing "good" medical studies from "bad" ones
- Understanding how scale relates to statistics
- Reading the news with a critical eye
- Understanding the facts about crime in the U.S.

This chapter is about numbers that are reported to you every day on radio and television news and in newspapers and magazines. It is not, I regret to report, written from an entirely neutral point of view, at least as television news depicts neutrality.

My point of view, which is supported by the numbers in the stories themselves, is that very often there is much less than meets the eye. Because the news carriers operate 24 hours a day, 365 days a year, they have built-in mechanisms for generating news stories from data that often doesn't really merit reporting. Have you, for example, ever seen headlines such as these?

- Three-Year Cholesterol Study Shows No Clear Results
- Most Crime Statistics Within Expected Bounds
- Police Work Safe Compared to Lumber Mill Jobs
- Stock Market Down for No Particular Reason
- Weather for Rest of Week Not Worth Reporting
- School System Doing About the Same
- Nearly All Gamblers Lose Money
- Most Small Commodities Investors Lose Their Shirts

I don't think that you have. Nobody wants to tell you this stuff, and very few people want to hear it. For example, working the night shift in a convenience store is significantly more dangerous than doing nearly any form of police work short of SWAT team assault. That's simply a matter of reading the statistics. Nonetheless, nobody has managed to put together a highly rated television show about the dangers of convenience-store clerking (there's a great movie called *Clerks,* though), whereas the thrills of police work are dramatized several times a night.

Similarly, I can tell you from decades of personal experience at the frontiers of biotechnology that most research programs, even well-organized, well-funded programs inspired by brilliant ideas, don't often result in real medical breakthroughs. If you are diagnosed with lung cancer, for example, the recommended course of treatment won't be much different from the one used 20 years ago, and your life expectancy from time of diagnosis won't be vastly better. That's not much to show for the amount of research activity devoted to this topic. For that matter, we don't have anywhere nearly enough to show for decades of anti-smoking ads. There is, however, no shortage of stories about medical progress and fantastic developments that are just a few years away.

This chapter is not meant to discount in any way the efforts of millions of fine people trying to do their best every day. It's a complaint about the way numbers get distorted in news to make all sorts of events seem more significant than they really are. So this chapter explains the "resizing" of numbers in the news so that you can get a better understanding of the meaning behind those numbers. If you get interested in this sort of analysis, find a copy of the book *200% Of Nothing* by A. K. Dewdney (John Wiley, 1993), a detailed and hilarious study of statistics in the news. (Actually, anything by Dewdney is worth reading.)

Warning! Read the Label Carefully

I picked up the paper yesterday and read an alarming story about coffee. According to the story, reported on the AP newswire, a researcher in the Netherlands had discovered that swallowing small amounts of coffee grounds resulted in a significant increase in serum cholesterol. The testing program involved drinking lots of Turkish coffee, which is pretty much like regular old Italian-restaurant espresso only with a dab of mud in the bottom.

So does that mean that it's panic time? Well, on the face of it, one reason not to panic is that you probably don't get much coffee with grounds in your diet. Practically all the coffee in the U.S. gets dripped through a filter, so you may wonder why this exact bit of news, if that's what it is, deserved a page-one, inch-high headline. But I picked this study because it has all the numerical earmarks of studies that you should generally ignore. Here's the list.

A small sample

Reading the article, you find that the study involved about 24 university students who were divided into two groups of 12. Essentially, one group got the Turkish coffee, and the other didn't.

It's hard to be polite about this. You could hardly prove that dogs bark with a sample size this small. The statistical power of a test is related to its sample size. In correctly designed tests, bigger sample sizes allow researchers to make more definite conclusions. You can treat yourself to a great simplification in your medical-news intake with the following easy-to-remember rule:

TIP

Don't bother reading medical results obtained on fewer than one thousand test subjects.

If you start noticing sample size in articles, you'll be amazed at the number of dramatic results reported from studies by a single doctor with 17, 31, or 43 patients. You may miss a few worthwhile results by using the preceding rule, but you probably won't miss much, because studies tend to get repeated if they're paying off. In fact, if someone obtains a really promising result on a smaller sample, getting funding to repeat the study on a large scale is usually pretty easy. You can, in most cases, wait with impunity until a bigger study either confirms or contradicts the original study results.

No controls

A good study has very tight controls on conditions — that means, for example, that only one factor is different between the two groups being compared. As you read along in this study, it develops that the two groups were matched for starting cholesterol levels and that some attempt was made to control diet. But if one of the students went off on a coffee-inspired ice cream binge during the test, the results of the ice cream on a sample this small would be almost enough to account for the whole effect observed in the test.

Lots of studies, for example, are performed on people who are taking some form of medication as they go about their daily lives. Later, researchers count on these peoples' memories of what they ate, what they did, whether they took aspirin, or any number of other things. Sometimes these other factors matter, and sometimes they don't. But this lack of control over other possible influences is one of the reasons that you can also ignore most studies that begin, "A doctor in Philadelphia reports that ordinary baking soda can cure herpes," or the equivalent.

My point is this: Always check to see whether the researcher has ruled out every other possible explanation of the results.

With a little reading, you will soon find that most news stories actually report at least some of the conditions of a test, including the way the test was designed to control for various factors that might be important. If you think about it carefully, you can often figure out for yourself whether all the possible influences in the test were really controlled.

Consider the coffee "experiment." Statistically, if you take a small batch of students and test their cholesterol levels today, split them into two smaller batches, and test again two weeks later, the most likely result is that you will find some difference in the two batches. Does this mean that something special happened? Typically, it doesn't.

The Effects of Scale

You may come across a medical report that has a decent sample size and effective procedural controls but that makes no mention of the effect of dosage (amount of a drug, amount of a treatment, amount of peanut butter). The absence of this information is quite damaging to your ability to understand what's going on, because there are typically two kinds of scale effect. You need to know which kind of scale effect applies to the study at hand.

Straight goods

Figure 5-1 shows one type of scale effect. For twice as much exposure to some factor, you get twice as much effect.

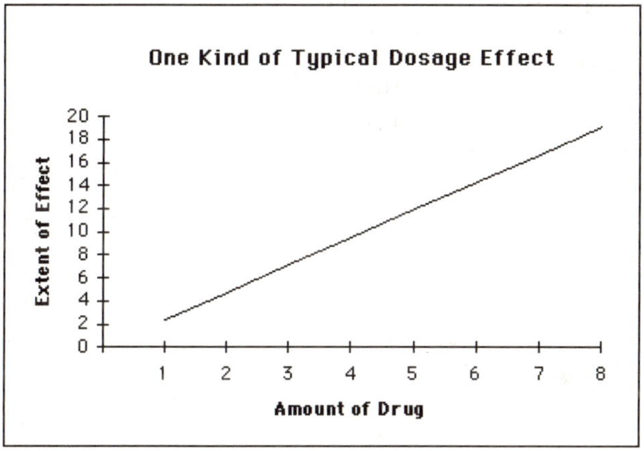

Figure 5-1: A simple dose-effect relationship.

One example of this kind of relationship is ionizing radiation and DNA damage. Sometimes this is expressed as, "There is no safe amount of radiation." Every exposure to x-rays, therefore, produces some measurable amount of DNA damage. If you get ten times the exposure, you get ten times the damage.

When rescue workers went into Chernobyl to try to contain the radiation leakage, they weren't taking a risk per se. In the interests of trying to stop a plume that would take out a hundred thousand people, they were going to absorb an amount of radiation that was absolutely guaranteed to kill them. And they knew it. You can look up the dosage in millirems (a measure of radiation) in a table and read off the result: certain fatality. There was no chance that they could "get lucky" or that the radiation would miss or that somehow a stronger personal constitution could overcome the damage.

Some types of chemicals have this same relationship to liver damage. At any amount, you get damage, and at twice the amount, you get twice as much. Sniff enough petrochemical solvents, and you're gone.

Dosage curves

Most effects in human biochemistry aren't like this, however. For most substances, particularly the everyday substances most likely to be mentioned in news articles of the "Whole Wheat Bread Causes Insanity" type, the dosage curve looks like Figure 5-2.

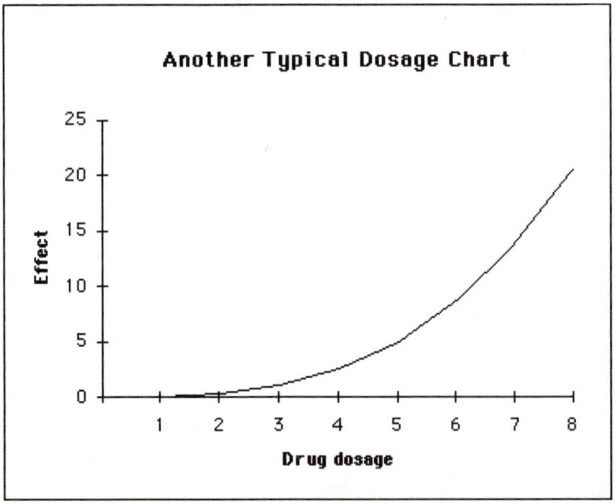

Figure 5-2: A more realistic dose-effect relationship.

What does this mean? Several things.

- If you drink a few glasses of wine per week, you'll be okay from a health perspective. If you drink a quart of Absolut Vodka every day for a few years, you're dead meat. It's not that drinking only 5 percent as much alcohol (as the quart-a-day schedule) has 5 percent of the effect — it doesn't have any adverse effect. The dosage effect is a curve, not a straight line.

- If you eat a half-pound of beef fat every day, you can acquire a whole variety of problems, one of which is rapidly inflating pants size (hey, that's a lot of extra calories). But eating small amounts of fat doesn't result in problems. Europeans visiting the United States are always startled to see what a collection of porkers we are, and grocery stores in Italy and France are not packed to the rafters with diet foods. You just eat much less of the same stuff, from potatoes to butter to beefsteak to ice cream, and it's all okay (as long as you get enough carbohydrates and fruits and vegetables).
- Remarkably, the health statistics for people who smoke only a few cigarettes (two to three) a day are barely distinguishable from the statistics for nonsmokers, especially in urban areas. In contrast, at three packs a day, you're up in the rapidly soaring danger zone of the curve, pretty certainly headed for a really ugly demise. The problem with this piece of information, and the main reason that medical authorities are reluctant to discuss it, is that experience shows that very, very few people hold on at the three-cigarettes-a-day level (and indeed, why bother?). But even cigarette damage obeys a curved dosage law rather than a straight line.

The conclusion from the two types of graphs is that you don't have the whole story until you know which type of graph is appropriate, and you don't know how to modify your behavior until you know the level at which some effect "kicks in." Badly calibrated results on different foods, for example, have made the U.S. a paradise for purveyors of goofy quack diets.

Crime

The good news is that you're not as likely to be shot by a criminal as you might think. The bad news is that virtually everything that comes out of your television is designed to keep you paralyzed with fear every day of your life. This section is intended to help you read crime statistics in newspapers (there isn't much in-depth statistical information on TV, I'm afraid) when you can find them. Of course, you are more likely to read accounts of individual lurid crimes than real statistics, but I will soldier on nonetheless.

Bang?

I used to live in Los Angeles, and I knew about five officers from the Los Angeles County Sheriff's Department personally. At a wedding reception one afternoon, I asked them how many shots they had fired on duty, in total (this was in the drug-crazed, crime-ridden late 1970s). The answer, to my surprise, was zero. As they explained, a long list of gun-handling rules have to be observed, and you can pretty much kiss your career good-bye if you violate them. The typical officer, not just in Anytown, Ohio, but in L.A., retires without anything resembling a gunfight on his or her record.

Contrast this, if you like, to the number of shots fired in a four-hour sampling in a single day on television. Day after day, year after year, the difference between real law enforcement on the streets and TV law enforcement gets wider (actually, the ancient Dragnet show was closer to reality), as pseudo-documentary TV shows scour the entire country trying to find instances of SWAT team live action to pep up the show.

Crimes of the century

One interesting and well-studied case of news crime statistics concerns serial murders. I'm just going to talk about serial murders here because there are several other statistical issues (incarceration rate, results of three-strikes-you're-out sentencing laws, or deterrent effects [or not] of the death penalty) that I wouldn't touch with an insulated fiberglass four-meter barge pole. After all, you should be able to read a math book without too many political hassles.

In the mid-1980s, many newspapers, repeating an original story that appeared in *The New York Times,* advanced the proposition that nearly 20 percent of all the murders committed in the U.S. were the work of serial killers. This number corresponded to about 4,000 murders, and the papers further suggested that this epidemic had exploded since the early 1970s. It was also suggested that half the victims were under 18 years in age and that the U.S. had about 40 such serial killers out wandering the U.S. at random, and pretty busily occupied at that. In the mid-1980s, the U.S. had about 5,500 unsolved murders each year, and a large fraction of these — nearly two-thirds — were attributed to serial killers.

These figures bring up a number of questions. First, the annual murder rate overall has been fairly stable. It doesn't double from year to year. Second, the number of murderers identified as serial killers from 1950 to 1970 was exactly two, and one of them was the notorious Boston Strangler. Did something very strange happen in the year 1970 to produce a large crop of demented maniacs?

Nah. We've always had about the same number of demented maniacs as a proportion of the population. Actual Justice Department figures say three things: 1) There may be about 30 serial killers in the U.S. at large at any time, 2) they typically stick close to home, and 3) they kill four to six people per year. That means fewer than 200 serial killer victims per year, plenty bad enough, but not much like the figure 4,000. And it means that probably less than 1 percent of murders are committed by serial killers, not 20 percent.

So what was going on here? What was the basis of dozens of lurid TV mock-documentaries? The explanation has to do with the way crimes are reported in the Uniform Crime Reports (UCR). The early 1980s saw an upswing in the number of crimes first reported as both "motiveless" and "no suspect" and that were subsequently listed as unsolved. Looking at an assortment of cities and noticing which ones had a big upswing in "unsolved" murders (in a really typical murder, the spouse is still standing there in the kitchen holding the gun when the cops show up), there was in fact a powerful connection between unsolved murders and drug-trafficking activity. It happened that, in the mid-1980s, most murders that involved finding the body of a known drug dealer in an alley, with neither drugs nor money nor witnesses about, was UCR-listed as motiveless and no-suspect. That means that a dead guy with half-a-dozen diamond rings and a matching number of small-caliber holes in his head, blown away inside a $200,000 sports car, was going on the books as "serial-killer victim," one more number to use to scare the daylights out of housewives in Joplin, Missouri.

The crude result of all this was that the original report of an upswing in UCR motiveless and no-suspect murders was directly translated into serial-killer murders in the widely publicized newspaper article, and that contents of that article were endlessly repeated. Any beat cop in Miami could have suggested that the dead bodies in the trunks of Ferraris were probably not the work of truly motiveless serial killers randomly attacking the public, but this perspective was never aired.

The math issue here is twofold. First, every time you read an article, you have to ask, "What is the basis of these numbers?" Second, it wouldn't hurt to ask whether the conclusions in the article violate common sense. This is pretty tough to do, since everyone's common sense is constantly being subverted by media distortions, but sometimes you can make a little investigation just on the numbers. In this case, the projected serial killers would have had to have been killing several people per week. Since there are no historical examples of real serial killers being able to maintain this rate (even Jack the Ripper averaged fewer than one per month or so), it is unlikely that some monster could be producing a few hundred dead bodies every few years in Cincinnati. Someone would notice.

A Bit More about Statistics

Tragically, statistics is regarded as a difficult subject. You gain almost enough ordinary experience in life to guess a great number of the most important statistical results, but you are almost certain to be put off by much statistics terminology.

Normal IQs and such

The curve in Figure 5-3 is called a Gaussian curve, or a normal distribution, and is sometimes informally called a bell curve.

This curve is characterized by a mean value (refer to the figure), which is the average point in the distribution. It's also characterized by a standard deviation, which is a measure of the width of the distribution. To take a concrete example, the normal distribution that fits standard IQ tests has a mean value of 100 (loosely, the average IQ is 100) and a standard deviation of 16 points. Because the curve has a mathematically defined shape, it's possible to find the area under it for different values of x by using the formula for the curve and integral calculus.

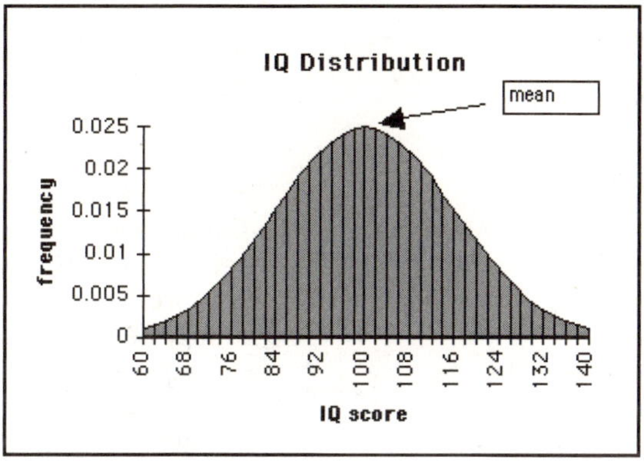

Figure 5-3: The normal distribution.

That, in turn, means that it's possible to determine the fraction of the population with IQs above some fixed number or below some fixed number. The organization Mensa, which tries to select members with high IQs, accepts test results that are two standard deviations above the mean, meaning 2 times 16 (32 points) above the mean of 100, or IQs above 132. It happens that about 2 percent of the population has IQs of 132 and above, implying that 98 percent of everybody has an IQ lower than 132. This definition (the top 2 percent) is usually the definition that high school counselors use for applying the term gifted, although nobody in the cognitive sciences would agree with that definition any more.

Normal piston rings and quality

This IQ business just happens to be one everyday use of the normal distribution. Another common use is in quality control. Suppose, for example, that you are manufacturing piston rings. If you carefully measure the diameter of a large set of piston rings, you will find a bit of manufacturing variation, as shown in Figure 5-4.

Here, the mean is 2.5 inches and the standard deviation is 0.0001 inch, or one ten-thousandth of an inch. With better manufacturing methods, you can aim to reduce this standard deviation, making the parts more strictly uniform. In fact, that's pretty much the philosophy of the industrial quality control methods perfected by the American statistician W. Edwards Deming.

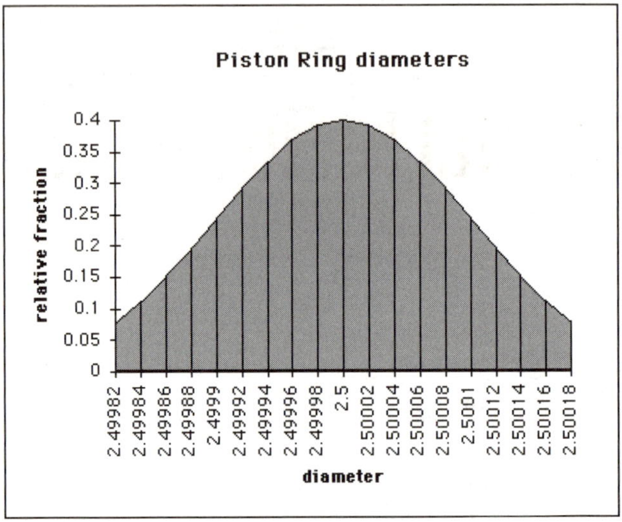

Figure 5-4: The normal distribution of piston rings.

Deming proposed that with tighter statistical specs, it would be easier to produce superior designs. He had a great influence on methods of production during World War II. After the war, in the Fabulous Fifties, American markets displayed a nearly limitless appetite for goods, and Deming's consulting services fell out of favor in the U.S. So he went to Japan and managed programs that turned Japan's output from a synonym for junk to the most rigorously quality-controlled manufactured goods in the world. The annual Japanese quality-control prize is called the Deming Award.

At one point in the 1970s, Ford Motor Co. bought a stake in the Japanese company Toyo Kogyo, which was supplying the small truck sold in the U.S. as the Ford Courier. A team from Ford visited Toyo Kogyo several times and finally asked, out of plain curiosity, who was the white-haired old gentleman (Deming looked a bit like Colonel Sanders of Kentucky Fried Chicken fame) in the pictures in everyone's offices at the Toyo Kogyo works. The Japanese executives couldn't believe that the Ford executives didn't recognize the famous American quality expert Deming. They may have believed it if they'd had an opportunity to drive my very own 1978 Ford Mustang II. From door handles to paint job to U-joints to electrical system, it was a veritable rolling laboratory in (lack of) quality control. Things are better now at Ford than they were in the '70s, I'm thrilled to report.

Chapter 6

General Math Shortcuts

In This Chapter

- Store math: Rounding and grouping prices
- Doubling and interest rates
- Checking your answers with the nines method
- Squaring numbers
- Multiplying not-square numbers
- Dealing with high finance on a calculator
- Converting and other metric matters
- Checking by round-off

This chapter, in my opinion, distinguishes this book from most other works on math. Needless to say, when I was first contacted about writing this book, I went out and bought everything even remotely comparable on the market.

What I discovered is that some of the calculating "tips" in rapid-math books really provoke the remark, "Huh?" The tip for rapidly multiplying a number by 12, listed in one such book for example, is as follows:

1. Multiply the number by ten.
2. Double the number.
3. Then add these two intermediate results together.

The author doesn't seem to have noticed that this is just the definition of multiplying by 12! It's not a tip, it's not a trick — it's exactly what they tell you to do in school. What I'm going to try to do in this chapter is at least to tell you something that you may not already know.

Store Math, Part One: Rounding

The people who run stores firmly believe that you perceive $17.99 to be significantly less than $18 — careful studies show that you tend to round the number down instead of up.

Of course, stores also use an elaborately calibrated class-system of price tag identification, ranging from fluorescent orange tags that say $7.00 to discreet little brass numbers that say 800 (no dollar sign). Sometime, take a field trip to some downtown store windows in a big city and watch how the stores signal their position in the retailing food chain with price tag conventions (besides just prices!).

Anyway, through the vast middle reaches of pricing, all this stuff about prices ending in nines reigns supreme. So here's a method for adding prices that also works with other numbers that can be rounded up or down easily.

Suppose the problem is to add these numbers:

EXAMPLE

$$\begin{array}{r} 13.99 \\ 7.98 \\ 21.99 \\ +\ 4.97 \\ \hline \end{array}$$

The last thing you want to do is spend five minutes carrying ones and so forth. So rewrite this in two columns:

$$\begin{array}{rr} 14.00 & 1 \\ 8.00 & 2 \\ 22.00 & 1 \\ +\ 5.00 & 3 \\ \hline \end{array}$$

What's going on here? To get the values in the second column, you just take the tiny difference between the starting number and a nice, even number with 00 at the end. You can think of the second column as a sort of correction factor.

Here's how you process the answer. First, add the 8 and 22 to get 30. Add 5 to the 30 to get 35. Then add 14 to the 35 and you have 49. So the total would be $49, except for the column of correction factors in pennies. 1 + 2 + 1 + 3 = 7, so the total of the original numbers is

$$\$49.00 - \$0.07 = \$48.93$$

Officially, you have done more mental operations by breaking up the problem this way, but they were all easier than the original addition would have been.

Store Math, Part Two: Grouping

In the preceding example, I slipped a little mini-trick past you in the addition. That is, I added 22 and 8 first instead of just doing things the textbook way. You will find very few addition problems in which judicious pairing of numbers for adding doesn't simplify things marvelously. That's especially true if the numbers have been conveniently rounded off first. Here's a nice, long addition problem. As one way to prove that this method works in general, I had a computer select all these numbers at random:

$$\begin{array}{r} 17 \\ 151 \\ 23 \\ 24 \\ 205 \\ 39 \\ 75 \\ \underline{66} \end{array}$$

First, look over the column and find all the pairs that make a sum ending in zero. If you add up all these pairs, you reduce the problem to

17	(17 + 23)	40
151		
23	(151 + 39)	190
24		
205	(205 + 75)	280
39		
75	(66 + 24)	90
66		

This third column doesn't have obvious groupings at this step, but you take (90 + 190) first to get 280 and then take (40 + 280) to get 320. These two values add up to the nice, round number 280 + 320 = 600.

The other advantage of doing this type of problem in pairing steps, even if you don't have absolutely perfectly convenient numbers, is that the answer is easier to check. You never add more than two numbers at a time, which puts less strain on your abilities, and you can check these two-at-a-time additions at a glance.

Playing Doubles: Interest Rates and Doubling Time

This section gives you just a little table relating interest rates to doubling time. The reason you may need to know this is that someday you may want to retire, which is going to take a considerable amount of money.

You'll need lots of money because of your fabulous extended life expectancy. When retirement age was first set at 65 in the Prussian State Civil Service in the late 1800s, the thrifty Prussians expected retirees to die within a few years of retirement, and the mortality tables confirmed this expectation. It was a simple business — you work for 45 years, save a little each year, and then spend it in two years of retirement and die before you have to see World War I.

You, on the other hand, are planning to work from age 21 or so until about age 65 and then live for about 20 more years. How much do you figure that you should save to work for 40 years and then retire for 20? Uh-oh. The Prussian arithmetic isn't going to work this time.

One simple way to calculate what your savings will be worth at some point in the future is to consider doubling times. Table 6-1 is a condensed financial table showing how long it takes for your money to double at different annual interest rates:

Table 6-1 Doubling Time versus Interest Rate

Annual Interest Rate	*Years to Double*
5%	14.2 years
7%	10.2 years
9%	8 years
11%	6.6 years

The problem, of course, is finding a reasonably safe investment that has a real return (adjusted for inflation) that's high enough — 9 percent or higher would be nice. That's beyond the first problem, which is collecting some money to invest in the first place.

Doubling for Multiplication

As long as the subject of doubling has come up, I may as well tell you about a doubling trick for multiplying numbers.

The procedure is based on the observation that any number can be made up of powers of the number two — that is, numbers made of multiplying two by itself. Look at this list of powers of two:

- 1
- 2
- 4
- 8
- 16
- 32
- 64
- 128

Any number less than 128 can be made up as a sum of those numbers from 1 to 64. Here are some examples:

43 = 32 + 8 + 2 + 1

55 = 32 + 16 + 4 + 2 + 1

17 = 16 + 1

So if you want to multiply 195 by 17, you can multiply it by 16 and then by 1 and then add the two results together. The only reason this rigmarole is worth any attention is that you can multiply a number by 16 or 32 or 64 and so forth just by doubling that number a certain number of times. The scheme for multiplying 195 by 17 goes like this:

1 □ 195

2 □ 390 (two times 195)

4 □ 780 (two times 390)

8 □ 1,560 (two times 780)

16 □ 3,120 (two times 1,560)

1 □ 195

and then adding

$$\begin{array}{r} 3{,}120 \\ +\ 195 \\ \hline 3{,}315 \end{array}$$

Again, the advantage of this method is that it's easy to check the steps. I also bring up this method here because it is essentially the way computers multiply numbers — the computer converts all the numbers into sums of powers of two and converts them back into familiar decimal numbers only when you have to look at them.

Using Nines to Check Your Answer

An old method for checking correctness of an answer to a multiplication or division problem is still worth knowing, in case you have to work without a calculator. The method is based on so-called "9s remainders," an ancient trick for checking calculations. To get a 9s remainder, you add the digits in a number and keep adding until you have a single digit. And if the digit is nine, the 9s remainder is said to be zero.

I'll just do some examples rather than give you a rule:

- **756:**

  Add the digits $7 + 5 + 6 = 18$.

  Add these digits $1 + 8 = 9$.

  9s remainder = 0.

- **631:**

  Add the digits $6 + 3 + 1 = 10$.

  Add the digits $1 + 0 = 1$.

  9s remainder = 1.

To check a multiplication, you find the 9s remainder for the numbers you are multiplying, and the product of the two numbers should have the same 9s remainder as these two remainders multiplied by each other. This isn't as bad as it sounds. Look at this case:

$$\begin{array}{r} 1{,}438 \\ \times\ 615 \\ \hline 7190 \\ 1438 \\ 8612 \\ \hline 884{,}370 \end{array}$$

Now, the 9s remainder of 1,438 is $1 + 4 + 3 + 8 = 16$, and $1 + 6 = 7$.

The 9s remainder of 615 is $6 + 1 + 5 = 12$, and $1 + 2 = 3$.

Multiply these to get $3 \times 7 = 21$; the check for this answer is $2 + 1 = 3$.

The rule says that you should then get 3 as the 9s remainder for the product 884,370:

$$8 + 8 + 4 + 3 + 7 + 0 = 30,\ 3 + 0 = 3$$

This seems like a bit of work, but it's a better alternative than just stepping through the multiplication again, where you are likely to make the same mistake you made the first time.

Squaring Numbers

Numbers ending in zero are easy to square. You just square the first digit and tack on a pair of zeroes. That is

$$30 \times 30 = 900$$

$$40 \times 40 = 1{,}600$$

$$70 \times 70 = 4{,}900$$

and so forth. Knowing a few squares can't hurt, especially since you get them for free when you memorize your times tables. The information comes in handy when you're deciding how many overpriced hand-made tiles you need for your kitchen floor, for example.

Multiplying Almost-Squares

What happens if you have a number that's almost, but not quite, easy to square? Suppose you have a number like 41. For 41, think to yourself 1,600 (the square of 40), add 80 (2×40) to this number to get 1,680, and then add 1 to get 1,681.

You're just following this rule from algebra:

$$(D + d)^2 = D^2 + 2Dd + d^2$$

In the case of 41, you're saying that 40 is the big *D* and 1 is the little *d,* and then you're plugging them into this simple expression. It looks like this:

$$(40 + 1)^2 = 402 + 2 \times 40 \times 1 + 12$$

$$(41)^2 = 1{,}600 + 80 + 1$$

High Finance

After you get your financial calculator, you will probably lose the little manual, so I'm going to review what the main keys mean. Photocopy this page about three times and file it in obscure desk drawers so that you'll always have a copy.

The main keys are the following:

- N
- PV
- FV
- %i
- PMT

N means *number of periods.* In a five-year auto loan, *N* is 60 because interest is applied monthly and 5 × 12 = 60. For a 30-year mortgage, the number is 360. Often people forget and just enter the number of years, which usually is incorrect.

PV stands for *present value,* but what it means for you is the amount you are borrowing. If you have an $11,000 car loan, PV is 11,000. If you have a $135,000 mortgage, PV is 135,000. Because most of the time you are figuring payments rather than calculating how much money you need to deposit to have some future value, this is the most appropriate working definition of PV.

FV is *future value*. You can enter a future value, an interest rate, and a number of periods and find out how much money you would need to deposit to end up with that FV after all the periods of interest. But you usually have the calculator figure out FV rather than entering the values yourself.

%i is *interest,* but it's the interest per period. That means that for most kinds of payments, you have to divide by 12 to get the number to use. An additional problem is that some calculators expect you to enter the number 8 if the interest rate is 8 percent, and others expect you to enter 0.08.

Here's how to check. Enter **15** for %i, **1** for N, and **100** for PV. Have the calculator compute FV (there's a key labeled CPT or Compute or something similar). If you get the number 115, then you can just enter the interest as the percent number — if the problem says 6%, you enter the number 6. If you get 1,600 for FV, then you should have entered **0.15** — the calculator wants you to enter **0.06** here if the problem says 6%.

And in problems where the interest is computed monthly (360 = N for the 30-year mortgage), you divide the interest rate (the 7% to 9% *annual* rate) by 12 before you assign it to %i.

PMT is *payment*. After you enter the other stuff, usually %i, N, and PV, you have the calculator compute the payment. If the payment looks about right, you entered the right values for %i and N. When you forget to divide by 12 on a monthly interest problem, the answers are way off, not just a little.

Metric Fun

Living as we do on this giant continent with miles of salt water separating us from everyone else, we get to use our own amusing medieval units for things. From time to time, however, you are obliged to convert things to units used by the rest of the planet. All you have to do to perform the conversion correctly is get the factor pointed the right way.

First point: Miles are bigger than kilometers. When you do a conversion, you end up with more kilometers than miles. The factor involved is 1.6. Now I'll explain what this means:

1) You have 138 miles. What's that in kilometers? You have to get more kilometers than miles, so you multiply by the factor

$$138 \text{ miles} \times 1.6 = 220.8 \text{ kilometers}$$

2) The sign outside Paris says 115 km (that's kilometers). How far away are you in real distance, Tex? The number for miles has to come out smaller, so you divide by 1.6.

$$115 \text{ kilometers} \div 1.6 = 71.9 \text{ miles}$$

With your foot flat on the accelerator of a Citroën-Maserati SM, you'll be there in a half hour!

Second point: Kilograms are bigger than pounds. When you do a conversion, you end up with more pounds than kilograms. The factor involved is 2.2. Now try some more conversions:

1) You weigh 165 pounds. What's that in kilograms? You have to have more pounds than kilograms, so you divide by the factor

$$165 \text{ pounds} \div 2.2 = 75 \text{ kilograms}$$

2) Your grandmother's German cookbook calls for 1.5 kilograms of bread chunks to make goose stuffing (remember to keep the schmaltz). What does this

mean in an American grocery store? The number for pounds has to be bigger, so you multiply by 2.2.

1.5 kilograms × 2.2 = 3.3 pounds

Rounding Numbers

Even if you use a calculator, you can mis-enter numbers. Probably the best way to check any arithmetic problem is to use rounded off numbers to make sure that you have the answer approximately right.

In the round-off check version of this problem:

$$\begin{array}{r} 1{,}438 \\ \times \quad 615 \\ \hline 884{,}370 \end{array}$$

I would take the first number as 1,500 and the second number as 600. That gives me four zeroes to account for in the end, and the multiplication of 15 × 6.

15 × 6 is just 90, and when I tack on the four zeroes I have as my rounded-off answer:

90 0000 = 900,000

Actually, this is within 2 percent of the correct answer. This problem is more complicated, in fact, than most. I would bother to keep two digits in one of the numbers in the multiplication only if the first digit is one. On a problem like the following:

$$\begin{array}{r} 4{,}458 \\ \times 3{,}129 \\ \hline \end{array}$$

I would just take the 4 and the 3 and try

4,458 approximates to 4,000

3,129 approximates to 3,000

4 × 3 followed by six zeroes = 12 000000 = 12,000,000

You can go through life and be pretty satisfied with your calculations if you always have the right number of decimal places.

Chapter 7

Using a Calculator

In This Chapter

- Choosing a calculator
- More on calculator accuracy
- Science students, beware: Determining whether your answers make sense
- Finding square roots
- Calculators and calories
- Having fun with the calculator alphabet
- Adapting recipes
- Working with sines
- Amazing your friends with numerical silliness
- Understanding reverse Polish notation

There are a great variety of calculators on the market, from simple four-function calculators (+, –, ×, ÷) that are given away free with a pack of ball-point pens to programmable scientific calculators that compete with small personal computers.

All calculators are not created equal. You may be shocked to know that they are not all equally accurate, but you may be pleasantly surprised to find that you can mimic many of the functions of fancier calculators on the cheapest four-function units available. Anyway, this chapter provides some considerations, tricks, gimmicks, and serious points for calculator users.

How Good Is Good Enough?

At the low end of the calculator food chain are lots of calculators that are accurate to eight digits. When you consider that these calculators are supposed to be used mostly as adding machines, this isn't a problem — you can keep track of dollars up to $99 million without a mistake. And, of course, once you get to the $100 million mark, you can afford to go out and spend $12 for a modest financial calculator, one that has keys for PMT (payment) and so forth. What the heck, live a little!

In the meanwhile, here's how to check out the numerical accuracy of the calculator you're using.

Divide the number 1 by the number 666 (not picked because it's beastly — it's just a good demonstration). You will probably see the following:

$$1 \div 666 = 0.0015015$$

Now, if you multiply this result by 666, you should get back to the number 1, right? More likely, you'll come up with this:

$$0.0015015 \times 666 = 0.999999$$

The reason is that the calculator would have had to carry at least two more digits (digits that don't appear in the eight-place display) to make a number that the calculator would "know" to display as 1.

You may think that this small discrepancy isn't so bad. In that case, try another example: 1 divided by 1,666.

$$1 \div 1{,}666 = 0.0006002$$

Now, when you multiply this last number by 1,666, you come up with this:

$$0.0006002 \times 1{,}666 = 0.9999332$$

Clearly, this answer is not the same thing as 1. The problem here is that to get this number to round off to 1 in the calculator display, the calculator would have to keep three more decimal places than it shows in the display (it would keep the extra digits in memory). It rounds off the answer to the eight digits in the display, and that accuracy isn't good enough to recover the original number.

What I said earlier about using calculators as adding machines still holds — you're not going to get addition errors. But for long series of multiplications, this sort of small error tends to accumulate. Every time you do a repeated division or multiplication, you get a little bit of error-creep, which in practice can easily make it into the third place after the decimal. For figuring mortgage payments to the penny, for example, you're better off buying a real financial calculator with keys for interest rates and payments than using a four-function calculator and plugging numbers into complicated formulas.

More on Accuracy

A scientific calculator, one with keys for sine, cosine, logarithm, and so forth, is usually more accurate than a four-function calculator. How many more decimal places beyond the digits in the display does a scientific calculator keep? Here's one way to find out. Call up π (pi) — nearly all scientific calculators have a key for π.

Try the sequence

$$\pi = 3.1415927 - 3.1415926536 =$$

Sometimes you get the result zero. Sometimes you get the result

6.0E –10

From scientific notation (the display with E in it), this translates to 0.0000000006. This means that despite the displayed number 3.1415927, the calculator really keeps the internal value 3.141592654. This is also true for many other "mystery" numbers, with the prime example being

$$e = 2.7182818$$

This number turns up over and over again in compound interest and other calculations based on growth formulas.

Science Students, Beware: Do Your Answers Make Sense?

I'm slipping this part in as a public service to my brave former colleagues laboring away teaching science in high schools and universities.

Here's the point:

TIP

Whenever you use your calculator, stop and think if the answer you are giving could actually

- Be possible
- Be measured

Is it possible?

I have seen students hand in tests in which they computed amounts of heat released in a chemical reaction that exceeded the heat of an H-bomb. In physics classes, it is not unheard of for students to find stones dropping from a building and subsequently exceeding the speed of sound on impact. This is the science class equivalent of calculating the mortgage payments on a $120,000 house and finding payments of either $2.17 or $53,789 per month.

When you see a problem on a test or even in a homework assignment that calls for a numerical answer, test your understanding of the problem by seeing whether you can write down what would be a reasonable range for the answer. A rock thrown from a tall building, for example, is going to hit the ground with a speed somewhere between 40 and a few hundred miles per hour (it's your job as a student to know what these numbers are in meters per second, too).

No matter what you find on your calculator, 0.24 miles per hour isn't a plausible answer, and neither is 24,000 miles per hour. If you get this point straight, you can have a great time on standardized tests like the SAT because the test designers usually include two ridiculous answers in each set of five. Immediately ruling out the ridiculous answers virtually guarantees that you will do well on such tests.

Is it measurable?

Another calculator problem that students encounter is reporting too many decimal places. Try this very basic physics problem:

A snail is in a desperate hurry to cross a driveway in the rain before a cat spots him. Therefore, he is hustling along at uncommon speed, covering 10 centimeters (cm) in 3 seconds. What is the snail's speed?

If you report the answer as 10/3 = 3.33333333 cm/sec, you have made an error that's almost as bad a reporting the snail's speed as 30 miles per hour. The data in the problem has only two decimal places, so you are justified in reporting the answer only as 3.3 cm/sec, no matter what the display on your calculator says. In fact, it would be a challenge to a well-equipped physics laboratory in a university to measure a speed like this to six decimal places, much less eight or nine.

Square Roots

Some plain four-function (+, –, ×, and ÷) calculators also have a √ key, but most of them don't. I just want to step through the guess-and-divide method for finding square roots because this is one of the few functions that turns up a lot in daily life. Every time you cut a piece of plywood on a diagonal or figure how much roofing material you'll need, a square root is lurking in the underbrush.

The rule

Here's the method: Look at the number for which you need a square root, say 8. Guess any number for a root (this first guess doesn't have to be good) — I could try 3. Divide it into the original number, giving $8/3 = 2.66$. Take the answer from this operation and average it with the earlier number (the average of 3 and 2.66 is 2.83). Repeat the process by taking this new average (2.83) as the starting point.

An example

For the sake of concreteness (or maybe I should say woodenness), I'm going to find the corner-to-corner diagonal length of a standard four-foot-by-eight-foot sheet of plywood.

One side of the sheet is

$$4 \text{ feet} = 4 \times 12 = 48 \text{ inches}$$

and the other side is

$$8 \text{ feet} = 8 \times 12 = 96 \text{ inches}$$

Looking at the picture in Figure 7-1 and invoking the Pythagorean theorem ($c2 = a2 + b2$), I have

$$\text{Diagonal2} = 482 + 962$$

$$\text{Diagonal2} = 2{,}304 + 9{,}216$$

$$= 11{,}520 \text{ inches2}$$

Now, I don't want inches squared — I want to know how long the diagonal is in inches, so I have to find the square root of 11,520. Also, because I'm just trying to get numbers in inches, I keep only one place after the decimal. If you've ever used a SkilSaw, you should be aware that tenths of an inch are a bit of a challenge to replicate in wood with construction-grade tools.

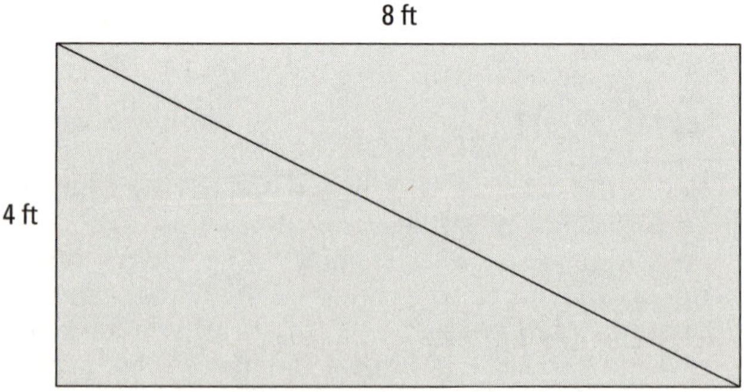

Plywood and Pythagoras: the diagonal length is the square root of 48^2 inches plus 96^2 inches

Figure 7-1: Plywood and the Pythagorean theorem.

To find the square root of a number such as 11,520, I just go through the steps that I just outlined, repeated here:

1. **Make a guess.**

   I'm going to guess 100, just because one side of the triangle was 96. That means that 100 has to be somewhere in the neighborhood.

2. **Divide the guess into the number you're trying to find the square root of (in this example, 11,520).**

   $$11{,}520 \div 100 = 115.2$$

   Actually, you could have done this without a calculator! Dividing by 100 just means moving the decimal point over two places.

3. **Average your first guess and the number that you find by dividing that guess into the original number.**

   In this example, my guess was 100 and I found the average to be 115.2, so

   $$\frac{(100 + 115.2)}{2} = 107.6$$

   That means that 107.6 is the new guess.

4. **Repeat the process.**

   That means to divide 107.6 into 11,520:

   $$11{,}520/107.6 = 107.1$$

Actually, the calculator says 107.06319, but I rounded it off to 107.1 because 0.06319 inches of plywood doesn't mean a lot to the average construction craftsperson.

The process "converged," as the mathematicians say, in only a few steps. That means that it produced a usable answer, and it produced it very quickly.

Food, Glorious Food

You probably can't believe that I'm going to bring up dieting after all this noise about square roots, but I am. This section is in here by request — my jolly study group in Sonoma County insisted that I include it.

Your body (he began portentously) is like a checkbook. That line would make a great introduction to a high school health movie, wouldn't it? You make deposits (food), you make withdrawals (activity), and you pay a sort of constant daily checking fee (resting metabolism).

To stay at a particular body weight, you need a certain number of calories from food, assuming a base of almost no physical activity. If you take in more calories than this "maintenance level," you start gently drifting up in weight until you wake up one morning in your early 40s about 30 pounds heavier than you were in college.

Your choices at that point are pretty basic: Take in less food or engage in more activity. Fairly active exercise — walking along modestly quickly — burns about 200 calories per hour. Really killing yourself is good for 500 to 600 calories. If you take in 3,500 calories less than maintenance level (or burn the equivalent in extra activity), you lose a pound. Here's the calculator problem:

Three cups of Haagen-Dazs butter pecan ice cream has 1,896 calories. According to the digital readout on the fabulous new exercise bicycle you just bought, setting 3 in "uphillness" is worth 580 calories per hour. How many hours do you have to move your sweating, sorry carcass on this fiendish device to work off the ice cream?

It's a simple division problem:

$$\text{hours} = 1{,}896 \div 580 = 3.27 \text{ hours}$$

Just to be on the safe side, may as well make it four hours. Hey, watch the movie Gettysburg while you pedal.

The same way you shouldn't make minimum payments on large credit card balances at 21 percent, you shouldn't eat real ice cream if you want to lose weight. Ice cream not a problem? A six-pack of Michelob is 960 calories. It's easier not to consume it than to work it off. It's just math.

By the way, one of the really annoying problems in human physiology is that your body adapts a bit to diets. Stop eating, and your calorie requirements drop somewhat. But they don't drop by a huge amount because you still need calories to maintain your body temperature and to operate your nervous system. The remarkable fact about your nervous system is that solving differential equations takes no more calories than watching Ren and Stimpy.

The Calculator Alphabet

The seven-segment characters in a calculator display lend themselves to an upside-down alphabet interpretation. The correspondence is as follows:

- B = 8
- E = 3
- G = 6
- H = 4
- I = 1
- L = 7
- O = 0
- S = 5
- Z = 2

Over the years, people have made up all sorts of entertaining and goofy puzzles using this letter code. Would I 1585 ÷ 5 to you?

Adapting Recipes for More or Fewer People

If you really, really have a hard time converting recipes — say a recipe for four people that you want to make into a recipe for six — you have two choices.

Choice one

Just give up on fractions. If the recipe says 5¼ cups, convert it to 5.25 cups. To scale this up to 6 people from 4, you just multiply as follows:

$$\text{quantity} = 5.25 \times (\text{new amount/old amount}) = 5.25 \times 6/4$$
$$= 7.875$$

You are unlikely to have measuring cups marked in decimal fractions, so you have to use a bit of sense. For example, 7.875 cups is just a bit under 8 cups (okay, it's 7⅞ cups). So just use a bit less than 8 cups. Any recipe worth using can survive at least 10 percent imprecision.

Choice two

For about ten dollars, you can get a calculator that does fractions directly. Look for a key that says

a b/c

You can then enter fractions like 3²⁄₇ directly and multiply them by 2⅕ if you like. The calculator reports the result in any fractional form you want. Problem solved!

Finding Sines

With a humble four-function calculator, you can do a pretty good job of approximating trigonometric functions.

Look at this short list with the angles in degrees:

- sin 0 = 0.000
- sin 15 = 0.259
- sin 30 = 0.5000
- sin 45 = 0.707
- sin 60 = 0.866
- sin 75 = 0.966
- sin 90 = 1.000

Even this short list is sufficient to let you interpolate sines of other angles. Suppose you needed the sine of 7.5 degrees. You can look in this list, find 0 for 0 degrees, 0.259 for 15 degrees, and guess that the sine of 7.5 would be halfway between. You're right, too. Getting sines correct to eight decimal places takes some ingenuity on the part of calculator manufacturers, but getting a "good enough" number to two decimal places is simple. I have never understood why some enterprising protractor manufacturer didn't simply print the sines and cosines of angles right there on the plastic — it would make the protractor useful in both geometry and trigonometry classes.

Numerical Silliness

Have someone write down a three-digit number, say 714, and then repeat it, writing 714,714.

Then have that person enter the number on a calculator, and explain that you can now identify factors of numbers (factors divide evenly into the number in question), among other feats of numerical prowess.

Stare at the number for a few seconds and then say, "Okay, right off the bat I can see that 7, 11, and 13 are factors of that number, and by the way, those numbers (7, 11, and 13) are all primes." The person with the calculator can confirm that these factors divide evenly into 714,714. Amazing!

What's going on here is that writing the number down in this way amounts to multiplying it by 1,001; and 7, 11, and 13 are the factors that make up 1,001.

Reverse Polish Notation

Calculators at the low end of the market follow the ordinary algebraic style of calculation entry. If you want to multiply 23 by 15, you enter 23, then press the × key, then enter 15, and then press the = key.

Hewlett-Packard, a long-time calculator pioneer, made and still makes calculators with a different entry style called Reverse Polish Notation (RPN). The name pays tribute to the work of a Polish logician named Lukasiewicz, but in fact this mode of entry is the direct descendant of the workings of old Friden mechanical calculators.

The keys for the preceding problem would be

23

[Enter]

15

×

in that order. You save a keystroke by doing the problem this way, and a lot of Hewlett-Packard's motivation in choosing RPN is that using RPN on programmable calculators enables you to pack fancier programs into a much smaller space. I bring up this topic just as a piece of calculator advice. HP calculators are nearly bullet proof (I have one from 1977 that still works!) and have tons of features, but the RPN models require that you make a small adaptation to your calculating methods.

Chapter 8

Credit Cards

In This Chapter

- Basic credit card math
- Payment plans and infinity
- Saving by borrowing?
- A plan for credit management

Credit cards are amazing. On the one hand, they offer lots of convenience and even bring some security, as we live in a large country where robbery for cash is far from unknown. On the other hand, they often represent the equivalent of a free financial piece of rope with a noose-tying video included in the package. The standard middle-class household can get itself into serious money trouble with credit cards in no time. Credit cards are the financial equivalent of pain-killers; sometimes they're just the prescription for you, but without careful attention, you can slide into an addiction that really complicates your life.

This chapter, of course, reviews the basic math involved in using credit cards. My real aim, however, is to get you to look at the faint gray type on the back of your credit card statements, review the tables in this chapter, and then throw yourself on the ground in a cold-sweat agony of abject fear.

By the way, please consult Eric Tyson's *Personal Finance For Dummies* for a coherent strategic credit management plan. My job here is to "run the numbers" with you. At the end of some calculations, however, you'll be able to formulate some advice of your own. The lessons in this chapter are about as complex as "Don't stick that silver fork in the electric outlet . . . a second and third time."

Understanding the Fine Print

I'm going to review a sample credit card statement, which is actually a sort of combined rehash of VISA, MasterCard, and Discover statements. The cards I'm discussing here are real credit cards, the cards designed to let you buy things that you may not actually be able to afford, not cards that force you to pay off your balance at the end of the month.

Making the monthlies

If you can handle the eye strain, the back of your credit card statement has a series of fascinating statements to interpret.

Statement 1: Figuring the minimum monthly payment

The minimum monthly payment is the greater of 2 percent of the outstanding balance and $10.

This means, of course, that if you owe $635.10 on the card, your minimum payment is

$$\text{payment} = \$635.10 \times 0.02 = \$12.70$$

If you owe less than $500, your minimum payment is $10.

Statement 2: Calculating finance charges

To compute the Periodic Finance Charges yourself, refer to the finance charge summary and use the following equation for each Average Daily Balance shown there:

$$\begin{aligned}&\text{Average Daily Balance}\\ &\times \text{Number of Days in Billing Period}\\ &\times \text{Daily Periodic Rate}\\ &= \text{Periodic Finance Charge}\end{aligned}$$

You may not be aware that the interest on credit card balances is typically compounded per day. Some credit card companies' accounting systems are fussier than others, with the computer noting the day of purchase and doing all this microcompounding on a day-by-day basis.

Also, you may be paying three different interest rates on different chunks of your overall balance. If your card allows cash advances, those usually accrue interest at the highest rate. There's also usually a difference between the interest charged on old balances versus that charged on the balance on purchases you made this month.

To simplify things, I'll just take the rate on unpaid, old balances. On the statements I'm using, the Daily Periodic Rate is 0.05425 percent. That makes an average monthly interest of

$$\text{monthly interest} = 30 \times 0.05425\% = 1.6275\%$$

That corresponds to an effective annual rate of nearly 21.4 percent if you start the year with an unpaid balance (although the card-issuing companies use the same numbers to arrive at a 19.8 percent rate by using a slightly different method for timing the compounding).

The paydown

As a first example, consider putting a balance of $2,000 on this credit card and making payments of $60 a month. This is a bit above the $40-per-month minimum payment the credit-issuing firm would want — I'll examine that particular case next.

The table in Figure 8-1 shows the results of the first set of payments. Basically, it's pretty simple. The card is charging about $30 a month in interest, and you're paying $60. Your payment covers your interest and makes a $30 bite into the balance.

Your balance thus drifts down fairly slowly. How slowly? Well, here's the bottom of the table (see Figure 8-2). Four years later, you make your last payment of a mere $25 or so (the last bit of the last month's payment is too small for $60). You have now paid about $900 in interest, too.

Here's a little algorithm (*algorithm* is the proper mathematical name for a trick). Remember this the next time you see a camcorder that you just can't live without. I'm not saying that you can't have the camcorder — I'm saying that you should know what it costs. And, lest you worry that you're being lectured by some know-it-all who has never done anything impulsive, you may readily guess that my morbid interest in credit card math is the result of some fairly bruising personal experiences. Here's the trick:

TIP

If you buy something on a high-interest credit card and don't pay off the balance at the end of the month, the real price goes up 50 percent.

That is, if you buy something on a card with an interest rate of 18 percent or higher, and you make payments that are about double the minimum monthly payment, the purchase will have cost you about 50 percent more than the original price by the time the smoke has cleared. Let me be specific:

- The big-screen TV for $1,799 really cost $2,700.
- The $80 shoes cost $120.
- The $165 anniversary dinner cost $247.
- The $2,300 used motorcycle cost $3,450.
- The $200 jacket marked down to $140 (30 percent off!) actually cost $210.

Making credit card payments: $60 payments on a $2000 balance

daily rate =	0.0005425
# of days =	30
factor =	0.016275
payment =	60

Month	Balance	Interest
	2000.00	32.55
1	1972.55	32.10
2	1944.65	31.65
3	1916.30	31.19
4	1887.49	30.72
5	1858.21	30.24
6	1828.45	29.76
7	1798.21	29.27
8	1767.48	28.77
9	1736.25	28.26
10	1704.51	27.74
11	1672.25	27.22
12	1639.47	26.68

Figure 8-1: Paying off a card at $60 per month.

Making credit card payments: $60 payments on a $2000 balance

daily rate =	0.0005425
# of days =	30
factor =	0.016275
payment =	60

Month	Balance	Interest
	2000.00	32.55
1	1972.55	32.10
2	1944.65	31.65
3	1916.30	31.19
4	1887.49	30.72
5	1858.21	30.24
6	1828.45	29.76
7	1798.21	29.27
8	1767.48	28.77
9	1736.25	28.26
10	1704.51	27.74
11	1672.25	27.22
12	1639.47	26.68
13	1606.15	26.14
14	1572.29	25.59
15	1537.88	25.03
16	1502.91	24.46
17	1467.37	23.88
18	1431.25	23.29
19	1394.54	22.70
20	1357.24	22.09
21	1319.33	21.47
22	1280.80	20.85
23	1241.65	20.21
24	1201.86	19.56
25	1161.42	18.90
26	1120.32	18.23
27	1078.55	17.55
28	1036.10	16.86
29	992.96	16.16
30	949.12	15.45
31	904.57	14.72
32	859.29	13.98
33	813.27	13.24
34	766.51	12.47
35	718.98	11.70
36	670.68	10.92
37	621.60	10.12
38	571.72	9.30
39	521.02	8.48
40	469.50	7.64
41	417.14	6.79
42	363.93	5.92
43	309.85	5.04
44	254.89	4.15
45	199.04	3.24
46	142.28	2.32
47	84.60	1.38
48	25.98	0.42

Figure 8-2: The end of the $60 monthly payments.

The preceding rule isn't accurate to three decimal places, but it's surprisingly accurate, and it's certainly good enough to remember as a guideline.

Faster Payoff

Having given you that gloomy bit of calculation, I will now consider the cheerier prospect of a faster payoff. If you make payments of $120 per month, you'll be all paid up after 19 months and will have paid a mere $351 in interest.

At payments of $180 per month, you're out of the woods in a year, having paid out $223 in interest. The table in Figure 8-3 shows the financial results for different schedules of credit card payback, and the sidebar entitled "Interesting" explains how to do this sort of computation yourself. The take-home lesson, however, is pretty blunt and requires no further arithmetic.

Paying off $2000 at 1.67% per month interest

Payment	Months	Total of Payments
$60.00	48	$2,905.96
$120.00	19	$2,351.09
$180.00	12	$2,223.30
$240.00	9	$2,166.67

Figure 8-3: Paying faster.

Historically (meaning way back in the '70s and '80s), paying off balances slowly was almost a viable strategy. Inflation was roaring out of control, and you would presumably make the last payments in nearly worthless funny money. In the meantime, you would have had four years' worth of pay raises.

Well, I say "almost" because even back then it wasn't a brilliant strategy. Now, of course, it's a screaming disaster. The inflation rate is a few percent, the "real" interest rate (the difference between the inflation rate and the prevailing interest rates) is at an all-time high, and it's been years since middle-class Americans have seen serious raises. You make a purchase for $2,000, and at a slow payback rate, you have to come up with about $3,000. And you are no more likely to have the $3,000 later than the $2,000 you would have needed to make the purchase outright. You can dig yourself a pretty big hole this way, and the sides of the hole are made of revoltingly slippery clay. By the way, this whole calculation assumes

Interesting

I did this computation in a computer spreadsheet for convenience, although you can certainly get the flavor of it with a simple calculator. You take the balance, apply the monthly interest rate, and then use the formula

Balance = Old Balance + Interest – Payment

This method of finding out how long you make payments is pretty tiresome, however. If you have a financial calculator, you can do it this way. Remember that the elements of the standard payment calculation are payment, interest rate, present value (the same thing as "amount" in this context), and number of payments. If you enter the other three quantities, the calculator can solve for the number of payments. Probably the simplest way to figure how much you pay at a particular payment level is to enter the amount, payment, and interest rate per month and just let the calculator return the number of payments. Multiply the number of payments by the value of the payments you made, and you see what you've shelled out for the duration.

that you haven't put any other purchases on the credit card in the meantime. If you tend to go trolling at the malls and throw a few more purchases on the total every weekend, the situation is nowhere near as rosy as I portray it here.

Making the Minimum Payments

It may be the case that you've been through a lot, you're tough as an old boot, and you don't scare easily. That means you're ready for the numerical equivalent of the famous shower scene in *Psycho.* It's called "making the minimum monthly payments." Your credit card calls for a minimum payment of 2 percent of the balance, or payments of at least $10.

For all I know, you who are reading this book are taking some kind of blood pressure medication, are prone to anxiety attacks, or suffer from insomnia. For these and for other technical reasons, I will present the minimum payment schedule in two separate tables, "Starting Out" (see Figure 8-4) and "Finishing" (see Figure 8-5).

In the beginning

Sit down somewhere, get yourself a cup of tea, and take a look.

Figure 8-4 shows the first year of payments; it's pretty straightforward. You pay out the minimum monthlies, which total $382.70. After paying this amount, you have made nearly $81 worth of headway into the original loan amount. That's a little more than $300 worth of interest and $81 worth of payments on the balance, just to beat you over the head with the point. Pretty brutal, huh?

465	104.88	1.71	10.00
466	96.59	1.57	10.00
467	88.16	1.43	10.00
468	79.59	1.30	10.00
469	70.89	1.15	10.00
470	62.04	1.01	10.00
471	53.05	0.86	10.00
472	43.91	0.71	10.00
473	34.62	0.56	10.00
474	25.18	0.41	10.00
475	15.59	0.25	10.00
476	5.84	0.10	10.00

Figure 8-4: One year of minimum payments.

Near the end(?)

It probably has occurred to you that this payment rate against the balance gets you nowhere fast. So let's look down this giant calculation to, oh, 30 – 40 years in the future. Way down yonder in the table, almost 40 years out, things have settled down a bit (see Figure 8-5). Every month, you make a minimum payment of about $10. But the interest on your balance is about $8.50. At a few dollars a month, you're not exactly streaking for the end zone. But don't worry, pretty soon you'll be making the $10 minimum payments and catching up a little faster. In fact, 40 years after the purchase, a mere $9,089 in payments later, you open the billing envelope and see that you have only $5.61 left to pay.

If you pay at this minimum payment rate, this book will be a yellowed chunk of carbohydrate resembling a squashed and mummified roll of paper towels before you finish paying for your original purchase. Your descendants will have to sell your original purchase as an antique to settle the debt at this payment rate. That's why, in the fine gray print on the back of your credit agreement, careful legal provisions call for immediate payment in the event of your untimely demise.

	Making credit card payments: $60 payments on a $2000 balance				
			daily rate =	0.0005425	
			# of days =	30	
			factor =	0.016275	
			payment =	60	
Month	**Balance**	**interest**			
465	104.88	1.71		10	
466	96.59	1.57		10	
467	88.16	1.43		10	
468	79.59	1.3		10	
469	70.89	1.15		10	
470	62.04	1.01		10	
471	53.05	0.86		10	
472	43.91	0.71		10	
473	34.62	0.56		10	
474	25.18	0.41		10	
475	15.59	0.25		10	
476	5.84	0.1		10	

Figure 8-5: Minimum payments, much later.

Please explain!

What's going on here? It's really pretty simple. The payments are a few cents less than the total interest per month. In that case, the payments never make a dent in the principal.

In the wild and woolly days of 12 percent inflation, the situation was sufficiently unsettled that most cards were geared toward complete payback within three years, even though credit card companies were inflamed by the prospect of collecting huge amounts of interest from you. Currently, as inflation is under control and credit-card interest rates are still (in most cases) fairly stratospheric, the companies are perfectly happy to leave the debt on the books and collect interest. You, as a credit card borrower, are the best investment they've ever seen. If you had a spare thousand dollars, there's nowhere you could park it where you would earn a 15 or 18 percent return on your investment, unless you had a crystal ball and an exceptionally compliant stock broker. But the credit card issuers are singing "I've Got You, Babe!" and merrily allowing you the easiest payments ever seen, except that they go on for eternity.

Chapter 9

Balancing Your Checkbook

In This Chapter

- Motivating yourself to maintain your balance correctly
- Easy steps for balancing checkbooks
- More checking details
- Approximation, or "good enough" checking

Here's some very good news: Doing a checkbook is easy. Begin with your balance, and from there, subtract what you spend and add in your deposits. The tricky part of managing a checkbook is balancing it (also called *reconciling your checkbook*), in which you make sure that your checkbook total matches the bank's total for your account.

What I'm going to tell you will very likely make the process of reconciling your checkbook easier. But the fact is that you can already do the arithmetic, especially with the help of a calculator. You don't have to be scared of the process.

Starting Simple

You want to avoid supporting your own bank in this grand manner. And it is relatively easy to avoid, too. You can keep your checking account in good order in just a fraction of the time you spend consulting the TV listings each month. That's why I propose to keep up the motivational approach in this chapter — you have the time; you own a calculator; and if you're bouncing checks, you're doing the equivalent of bouncing them on purpose. And it never pays to do so, no matter how broke you are. The bank makes sure of that.

Look at Figure 9-1, which shows a blank sample page of a checkbook register.

			Transaction	Deposit	Balance
Check No	Date	Transactions	Amount	Amount	

Figure 9-1: A sample blank checkbook page. Look familiar?

Some people use every line of the checkbook register to record actual checks. The preferred style is to use the gray zones for information, such as the purpose of the check (although going to a Safeway once a week and writing groceries in the gray space under every Safeway check is a bit ridiculous). Another tactic is to use the gray spaces for deposits rather than checks so that you can scan down the page and locate deposits quickly.

In real life, the gray spaces can be pretty useful as well as plentiful. When you get a standard box of checks and a check register, the register has enough space for a white and gray line per check, with room left over for your life story.

Figure 9-2 shows the simplest case imaginable — a brand new checking account with only a few checks written.

Notice also that to make things simpler, I have used unrealistically rounded-off numbers for each check amount.

Now, inspired by the information that almost 15 percent of adults in America have never, not once, reconciled their checkbooks, I ask you to do the totals right here, writing the numbers in pencil right in this book. See? Nothing to it.

Check No	Date	Transactions	Transaction Amount	Deposit Amount	Balance
					500.00
101	9/1	Ramos Shoe Repair	20.00		
102	9/2	Safeway	110.00		
103	9/5	Time-Life Books	35.00		

Figure 9-2: Checking: a basic example.

My guess is that people avoid total nonstop disaster by inquiring about balances at an automated teller machine (ATM) or by calling the bank almost daily. Please keep in mind, however, that the number you are given at the ATM or over the phone isn't your real balance since it doesn't include checks you have written that haven't made it back to your bank yet. Actually, some people get to be quite skilled at guessing balances, but doing so involves an unconscionable amount of background worrying. So please write in the totals for this example to get some practice. Figure 9-3 shows the solution.

Basic Checkbook Arithmetic

In the preceding example, you had two things in your favor. First, you had a starting point. It said right at the top of the page that the starting balance was $500, and the first listed check, 101, was the first ever drawn on this account. Second, the numbers themselves were easy. If you got paid $4,000 net income per month and all the checks you wrote were multiples of $100, checking wouldn't present much of a problem.

Check No	Date	Transactions	Transaction Amount	Deposit Amount	Balance
					500.00
101	9/1	Ramos Shoe Repair	20.00		480.00
102	9/2	Safeway	110.00		370.00
103	9/5	Time-Life Books	35.00		335.00

Figure 9-3: A checkbook in fine order.

In Figure 9-4, you see where it all goes wrong in the world of checking. The top of the page contains a total, all right, but there's a daunting arithmetic problem. A huge fraction of the checkbooks in this great land of ours look just like this — checks pretty faithfully recorded, without a running total anywhere in sight. Another large fraction of checkbooks look like this example, only without a total at the top of the page.

Check No	Date	Transactions	Transaction Amount	Deposit Amount	Balance
					500.00
101	9/1	Ramos Shoe Repair	17.58		
102	9/2	Safeway	89.53		
103	9/5	Time-Life Books	24.97		
104	9/5	Electronics Today	17.95		
105	9/5	Pacific Electric	71.81		
106	9/5	Brite Cleaners	15.80		
107	9/6	Southside Saloon	39.00		
108	9/7	Home Hospice Foundation	40.00		
109	9/7	(cash at ATM)	80.00		
110	9/8	Molsberry's Market	66.48		

Figure 9-4: A checkbook with some real numbers, including sales tax.

The problem is that once you skip the subtraction on just a few entries, you aren't going to catch up unless you sit down at the old kitchen table and make it happen sometime at night. That's right, I'm hallucinating.

Various fix-ups have been offered through the years, of which checkbook calculators probably made the most sense (but if you remembered to use the calculator, you could do your math right there in the store).

A Sporting Proposition

Look down the list of numbers in Figure 9-4, and look at the total at the top. Now stroll with me (I've given you this account!) over to an ATM cash machine located at the end of the block. The question I have for you is simple:

Do you think you can get $40 instant cash?

Well, do you feel lucky, punk? Want to bet me $100 that the transaction will go through? Oddly enough, I have been in this situation dozens of times, wandering off with someone to a cash machine only to find that my companion really didn't know his or her account balance with any certainty, often with embarrassing results.

Figure 9-5 gives the actual total for these numbers. But I'm writing this book, so of course I already calculated the answer. I wouldn't take your $100 under these unfair circumstances.

			Transaction	Deposit	Balance
Check No	Date	Transactions	Amount	Amount	500.00
101	9/1	Ramos Shoe Repair	17.58		
102	9/2	Safeway	89.53		
103	9/5	Time-Life Books	24.97		
104	9/5	Electronics Today	17.95		
105	9/5	Pacific Electric	71.81		
106	9/5	Brite Cleaners	15.80		
107	9/6	Southside Saloon	39.00		
108	9/7	Home Hospice Foundation	40.00		
109	9/7	(cash at ATM)	80.00		
110	9/8	Molsberry's Market	66.48		**36.88**

Figure 9-5: A $40 near-miss.

The $40 withdrawal is just going to miss. At least at an ATM, you can find this out for free. If you write a check for $40 at this point, the bank will have a party at your expense. You may have an account that makes automatic transfers, charging you for the privilege. Or the bank may just go ahead and bounce the check and let the merchant to whom you wrote it charge you for the returned check while using it for cash register decoration, to the amazement (or amusement) of your friends.

Reality Time

You don't want this nasty bounced-check business to happen to you. I don't want it to happen to you, and our only connection is through this book. So it's time to examine your own checking account to see what needs fixing.

The main problem is that if you have let things slide, you don't have a starting point for reconciling your account. If you had a starting number, you could do all the subtractions necessary to get a balance — assuming that you had been recording your checks.

Unfortunately, you don't have a real balance, and the bank doesn't either. The bank shows a current balance, but it doesn't know how many checks you have floating around in the mail or elsewhere, so its balance is no better than your guess. The catch is, of course, that the bank has the last word.

Getting to the point

You can get to a starting point in several ways:

- If you're lucky, your bank prints an instant statement at an ATM. This isn't the best starting point, however, since it usually reports a fairly limited number of recent checks.
- Your bank probably provides a phone service that lets you call and see what the bank is reporting for your balance and also find out which checks have cleared. If you can use this service for a long enough history of checks, it works pretty well.
- At the end of the month, you get a bank statement with a listing of a month's worth of checks. This is probably your best bet for finding a starting point.

The monthly statement is the best starting point if you're completely lost because it includes a long list of checks. The lurking joker in the deck is always the check that you wrote two months ago that hasn't been posted yet at your bank. My own personal record is a check for $175.19 to a bookstore that took three months to clear, probably because the store owner went out and partied all night after finally unloading the three-volume *Dictionary of Byzantine Civilization* on someone and forgot the check in the pocket of his tuxedo. But if you have lost track of the real checking total, you can't get a true balance unless you're sure that all your old checks have cleared.

What the statement means

Your bank statement has two parts (a front and a back), not only because it's a two-dimensional piece of paper but also because it lists two kinds of information. On the front of the statement is a listing of checks cashed prior to the date listed on the statement. Figure 9-6 shows a little segment of such a listing. Notice that one of the items in the listing has an asterisk next to it, indicating that the number sequence of checks has been interrupted.

CHECK	DATE PAID	AMOUNT
766	11/15	36.00
767	11/15	31.46
770*	11/16	115.12
771	11/16	58.38
772	11/17	61.50

Figure 9-6: Checks out of numerical order.

There are three possibilities at this point:

- A check got destroyed (you may have written it incorrectly and torn it up in the store).
- The check has already cleared and appeared on last month's statement. If you have that statement handy (that is, properly filed in the correct shoe box), you can check. If you look in your own check register and see that the check was written to a local place that typically cashes checks at your bank the very next day, the check probably has cleared already.

- The check is flying through never-never land but will indeed land someday. You need to consult your check register for the amount.

On the back of your monthly bank statements, there are usually some well-intentioned forms for reconciling your account. They are about as inviting as your W-4 statement from work or the front page of your income tax forms.

Figure 9-7 shows what the bank is trying to tell you.

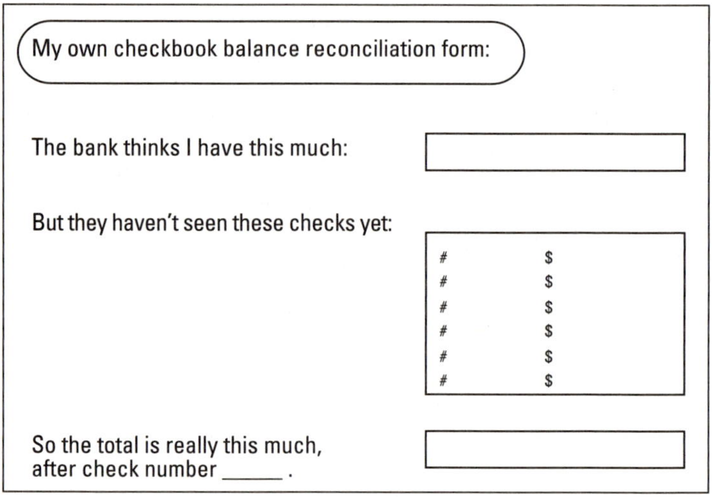
My own checkbook balance reconciliation form:

The bank thinks I have this much:

But they haven't seen these checks yet:

$
$
$
$
$
$

So the total is really this much,
after check number _____ .

Figure 9-7: A reality-oriented reconciliation form.

The bank is saying to you, "Look. As of the last check we show on the front of this document, we think that you have this much. We have taken out our monthly fee. We have also subtracted all those little charges for using an ATM at a different bank. You may have forgotten about them, but we never do. The only thing missing for a correct balance is that we obviously don't know anything about missing checks or checks written after the statement date."

So here's what you do.

Reconciliation part one: up to the statement date

1. **Take the bank's total and write down the last check number. Put each number in the right space.**

2. **Look back at the bank's check listing and note which checks are missing (an asterisk or some other indicator will tell you when the number sequence has been interrupted).**

3. **Decide whether the checks missing in the bank's list have already been cashed, are really missing (destroyed), or just haven't landed yet (that is, by statement time).**

   The surefire way to resolve this is to keep all your old bank statements since, statistically, the first case is the most likely.

4. **If the checks missing from the list really are outstanding, write them in the little spaces.**

5. **Get out your calculator and do the subtraction.**

   Usually, there will be only a handful of checks that you need to track down to make this adjustment (see Figure 9-8).

My own checkbook balance reconciliation form:

The bank thinks I have this much: $603.51

But they haven't seen these checks yet:

#455	$ 24.30
#517	$ 146.00
#518	$ 31.15
#519	$ 21.87
#520	$ 59.03
#	$

So the total is really this much, after check number #520: $321.16

Figure 9-8: Typical statement balancing involves just a few checks.

Congratulations! Following these steps brings you into agreement with the bank for the statement date. Doing so also gives you a starting point for next month.

By the way, if you shop for bargains, you will find that you can get a printing calculator with the charges you save from avoiding just one bounced check. If you really, really hate this sort of activity, it helps to buy a calculator that you actually like (maybe you think it looks cool) as motivation. Of course, the uptown way to do any of this is to get a copy of the computer program Quicken or, alternatively, to do your calculations in a spreadsheet. But if you haven't bothered to do your totals for months and hate this stuff anyway, I bet I can't talk you into a computer solution for your checking problems.

Also, reading your statements from time to time is quite informative. You may find that all sorts of funny little charges are applied to your account. One of the strangest categories is nonbank ATM charges. For example, do you get charged a fixed amount for using your ATM card at the ATM unit at your supermarket checkout? If so, what's your motivation for not writing a check?

In the old days when there were just checks, you didn't have to keep track of all the little bank-bites ranging from 20 cents to two dollars or more — another reason that keeping an account correct to the penny is fairly hard to do. One more reason is direct deposit of payments, which sometimes toggles by small amounts with changes in tax law. By using the statement at the end of the month, you get a starting point with the actual correct-to-the-penny amount. (It's like Hansel and Gretel finding the trail of bread crumbs in the woods.)

Reconciliation part two: up to right now

You now have a check number and a correct balance. That is, after that last check cleared at the bank, you had the balance that you just calculated. Please note that the statement date is probably a week old. It's time to bring you up to date with your most recent check.

1. **Using the same form (go make a bunch of photocopies of Figure 9-7 before you use it!), make a note of your starting balance.**
2. **If you made any deposits, add those to the balance and enter that number in the box.**

3. **Now go look up all the checks that you have written since the last check on the bank statement. Add them all up to get a single number total. Put that number in the checks column and subtract.**

There you are. You've got a total! It's time for a little heart-to-heart talk.

One thing you can see from this exercise is that if you fail to write down checks, there's no way to find your balance. The bank will get all your checks sooner or later, but you'll never know what's going on yourself at any given moment. As far as that goes, I recommend printing the check transactions instead of doing them in standard handwriting. It's just a little habit that makes you write the numbers more neatly.

Also, I know that many times people don't check for a total simply from fear. The problem is that whether or not you want to know if you're in trouble, the numbers won't forgive you. You're better off knowing that you're going to be $50 in the hole a few days ahead of time instead of getting a little letter from the bank. Maybe you can borrow a bit from another account, spend an evening rolling quarters, or hold off on a payment that would put you into bounce mode. Anything beats forking over a pile of overdraft charges to the bank.

The Last, Best Hope for Checkbook Problem Children

The problem with the first sections of this chapter is that you could have taken all that advice anytime. There must be 20 well-intentioned magazine articles a year about checkbook balancing. The back of your bank statement tells you the same thing. You can buy computer programs that basically play back to you the same set of steps that everyone else tells you to carry out on paper.

I would modestly suggest that I have outlined the simplest version of this process ever to see daylight, but . . .

You're still not doing it, right??!!

I'm not trying to be insulting — I'm your friend here. But you know who you are. Check for the warning signs:

- You've got a checkbook register in which the balance column never has a balance.
- Every three months or so, you walk up to an ATM and ask it for money, and it tells you that you have –$15.91 in your checking account.
- You go away for a weekend, and when you come home, you find six identical little envelopes from your bank in the Saturday mail, each one telling the same tiny sad story about the bank covering a piddling overdraft and charging you $10 or $20 to do it.

Not a pretty picture. I feel your pain. It used to happen to me, too.

And the problem won't go away by itself. I pointed out to a friend of mine that his overdraft charges in 1994 alone would have supported a trip to Hawaii — and rather than ask me for helpful checkbook advice, he encouraged me nevermore to dwell upon this possibility (imagine!). What I'm going to suggest is that you do a correct checkbook reconciliation from time to time and then keep accurate day-to-day track with a simpler arithmetic procedure.

TIP

Rounding for the cautious

If abandoning the idea of an exact total really makes you nervous, just round the numbers up every time. Even $11.04 gets rounded up to $12. It's not as accurate as the real rounding procedure, but it's got insurance built into it. You do less arithmetic, and your checking account always has just a bit more money in it than your account shows. The problem is that you don't know how much more, and you might start feeling optimistic. Remember, you're reading this chapter because optimism may have caused problems in the past ("Yeah, I'm pretty sure that check should be good . . .").

Always rounding up to whole dollars is the equivalent of establishing a miniature secret savings account into which, on the average, you deposit $.50 every time you write a check. After 40 checks, you have a built-in cushion of $20!

Don't lose hope. There's a way out of this problem, but it's a radical departure from the standard advice for checkbook management. For the rest of this chapter, I'm going to assume that you will probably never again have a checkbook balance that's correct to the exact penny. But you won't care. You won't get into trouble, either, and if you bounce a check, you'll at least be doing it on purpose instead of by accident. This is a last-resort, only-if-you-have-to procedure, but it works.

Pennies? Who needs pennies?

There is an easier way, and it involves accepting a slight loss of accuracy. How much accuracy? You get to decide. If you take one of the checkbook pages presented earlier (refer to Figure 9-5) and simply round off every entry to an even number of dollars, you get a balance of $36 instead of $36.88. The rounding procedure is simple: If the cents part of the figure is $.50 or less, you count it as zero, and if it's $.51 or more, you round up to the next dollar. After all this rounding, notice that the total is within a dollar of the correct figure.

TIP

The point, for this purpose, is to notice that if you keep track of your checkbook just in terms of whole dollars, your recorded balance will be very close to the real balance as long as you bring it back in line every few months.

This trick works because the differences between the rounded numbers and the true numbers cancel — about half the differences are positive and half are negative. If you want to be extra careful about never bouncing a check, deduct an extra dollar every ten checks or so.

If you don't keep track of pennies in your checkbook, you cut your arithmetic workload almost in half. Hang on — I have an even more radical proposition than just dropping pennies. I'm going to keep cutting the arithmetic workload until you're willing to do the numbers right there in the supermarket line.

Radical rounding

If you can handle it, you can take an even more radical approach to rounding. Try rounding to the nearest $5 — some people like doing so because years of making change with dimes and nickels and quarters seems to make people better at arithmetic problems where all the numbers end in five or zero. Figure 9-9 shows the results.

			Transaction	Deposit	Balance
Check No	Date	Transactions	Amount	Amount	500.00
101	9/1	Ramos Shoe Repair --17.58	20.00		480.00
102	9/2	Safeway -- 89.53	90.00		390.00
103	9/5	Time-Life Books -- 24.97	25.00		365.00
104	9/5	Electronics Today -- 17.95	20.00		345.00
105	9/5	Pacific Electric -- 71.81	70.00		275.00
106	9/5	Brite Cleaners -- 15.80	15.00		260.00
107	9/6	Southside Saloon -- 39.00	40.00		220.00
108	9/7	Home Hospice Foundation --40	40.00		180.00
109	9/7	(cash at ATM) --80	80.00		100.00
110	9/8	Molsberry's Market -- 66.48	65.00		**35.00**

Figure 9-9: Rounding checks to the nearest $5.

You write the real figures in the checkbook (writing them in the gray area is probably easier) and write the rounded figures in the Transaction Amount column. Note that, following this trick, you get $35 instead of $36.88. Given that you probably didn't know what all the other little charges from the bank were this month, this level of accuracy is not bad. To be extra careful, charge yourself an additional $5 after every 20 checks. (Technically, this trick prevents random errors from getting out of hand.)

So what happens if you round to the nearest $10? Amazingly (at least until you think about it for a while), this rounding procedure gives perfectly acceptable accuracy on a month-to-month basis. Look at the results in Figure 9-10.

So how does this sound to you? You get your checkbook in order once every two or three months to give yourself a starting point. After that, you only have to work with subtractions involving ten dollars. Those earrings for $49.23 after tax . . . 50 bucks. A check for $4.22 . . . zero. Now when you're standing in line at the supermarket and the total is $89.17, you will be subtracting $90 from, say, $400. Hey, you can do that one while the whole world waits and watches. Tell 'em to look at the tabloid article about Madonna bearing Barney the Dinosaur's love child to pass the time while you casually subtract a single digit right on the spot.

			Transaction	Deposit	Balance
Check No	Date	Transactions	Amount	Amount	500.00
101	9/1	Ramos Shoe Repair --17.58	20.00		480.00
102	9/2	Safeway -- 89.53	90.00		390.00
103	9/5	Time-Life Books -- 24.97	20.00		370.00
104	9/5	Electronics Today -- 17.95	20.00		350.00
105	9/5	Pacific Electric -- 71.81	70.00		280.00
106	9/5	Brite Cleaners -- 15.80	20.00		260.00
107	9/6	Southside Saloon -- 39.00	40.00		220.00
108	9/7	Home Hospice Foundation --40	40.00		180.00
109	9/7	(cash at ATM) --80	80.00		100.00
110	9/8	Molsberry's Market -- 66.48	70.00		**30.00**

Figure 9-10: Rounding to tens.

"But," you may object, "this is crazy! It's just too weird rounding off numbers this way." Well, no, it's not; the procedure works, and you don't need a calculator. If you do a proper reconciliation every 50 checks or so, you'll very seldom be more than $25 lower than your own records show, half the time you'll have a bit more money than you think, and typically you'll be within $5 of the exact balance. You can also just charge yourself $10 every 20 checks for using this procedure, just to be safe.

Isn't that better than being totally lost most of the time? Isn't it better than being $132 out of whack because you never do the subtractions? Doesn't it beat getting all those smirking little notices from the bank as it pockets outrageous fees? If you keep a checkbook register like Figure 9-10, you'll have all the correct documentation, you'll have the advantage of arithmetic you can do right when you write your checks, and you'll always show a balance that's "good enough" so that you don't get into trouble.

Actually, I have another motivation here. If I can recast the arithmetic so that you're willing to do it, the whole subject of checkbook balance reconciliation becomes a sort of game instead of a chore. If you know that you're within a few dollars of the correct balance, you'll want to see what the real to-the-penny total is when your bank statement comes.

What about Quicken?

If you at least have the patience to type in all the entries in your checkbook into a computer-based checkbook manager, you can get the arithmetic done for you automatically. The program Quicken from Intuit Corp. is available for PCs and Macintoshes and is the clear market leader.

A Quicken screen (see Figure 9-11) looks pretty much like a check register page, but it can do all sorts of other tricks. It can pay bills automatically with electronic bill-paying services, link your checkbook to credit cards, and pass all your financial information along to income tax software. It's also quite easy to learn.

The only catch is that you have to be willing to enter the information — if you don't keep Quicken up to date, it can't help you. But if you are willing to make the minimal effort of typing in the data (probably 30 checks a month in most cases), Quicken will take care of all your checkbook balancing and other transactions.

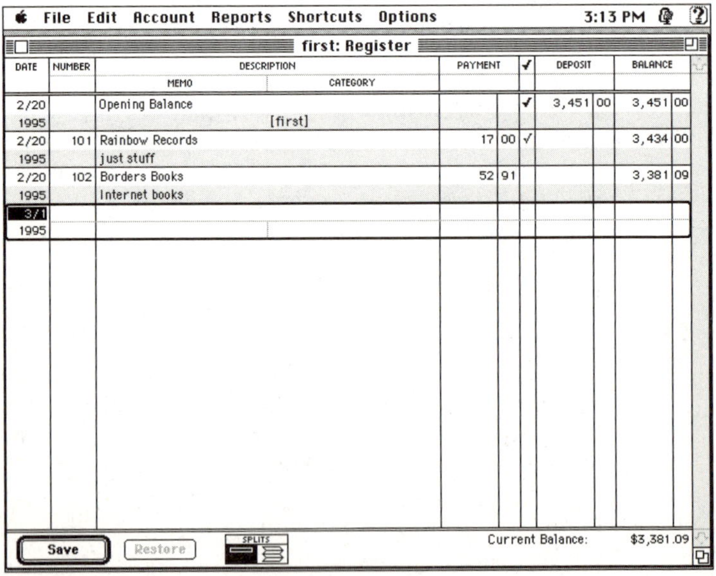

Figure 9-11: A sample Quicken screen.

Chapter 10

How Classroom Math Connects to Business Math

In This Chapter

- A few bitter remarks about high school math
- Sums in algebra, sums in money
- Geometry on plywood
- Thinking about risk

A good high school algebra course prepares you to do very well on the SAT test so that you can get into the college of your choice. It also helps you deal with problems involving trains headed toward each other at different speeds and their subsequent fates. It generally fails, however, in three key areas:

- Most of the work with formulas that you'll do in later life will be in setting up expressions that a computer will use. These expressions may be in a real programming language such as C or BASIC, but you may also have to deal with formulas in computer spreadsheets such as Lotus 1-2-3 or Microsoft Excel. Your algebra class probably doesn't offer training in anything like this.
- You could figure out all the financial math in this book for yourself if someone told you how to use the y^x key on the scientific calculator that you bought for algebra class. My guess is that nobody will. Most textbooks still treat calculators as an afterthought, with a few remarks relegated to an appendix.

- All sorts of fascinating developments in computer algebra have been roaring along. Not only is this material interesting in its own right, but a computer-accelerated course in algebra could get to advanced topics (group theory and modern algebraic structures, for example) in a one-year course. It's highly unlikely that any of this gets mentioned in a regular high school algebra class.

And just to be as direct as possible about the shortcomings of traditional algebra course content, it's quite typical that even the better students aren't able to make basic financing decisions later. They can solve lots of goofy artificial problems about trains, but they can't tell when changes in interest rates mean that they should think about a mortgage refinance, and they believe amazing and fantastically contradictory probability results.

Business Meets Science on a Calculator

Although many of the things you learn in algebra are of importance mostly in physics and in subsequent math courses, some bits of algebra impact you every day.

Breaking even

In the endless quest to find some activity that makes money, you are constantly confronted with possibilities in which start-up costs and revenues are based on activities. These are just a few examples:

- You own a small graphics shop and are thinking of expanding your services. Should you buy a color copier? How many copies would you have to sell, and at what price, to make the copier pay for itself?
- You could buy a used truck and start a parcel delivery service. How much business would you have to do per week to pay for the truck in the first year?
- How many silk roses at $5 apiece do you have to sell as an annoying mime at a Renaissance festival vending booth to pay back the $120 that you spent on materials and the $30 vendor's fee?

You can solve these problems by using direct algebra or by graphing the solutions, which gives you a nice picture of the break-even points in these problems. Figure 10-1 shows the break-even point for the first example. A graphical solution makes a strong argument in a meeting, even if most of the people at the meeting can't follow simple algebra.

A huge fraction of all the computer spreadsheets humming away on the desktops of corporate America are figuring out break-even points for models that keep track of dozens of factors. Similar models are used to optimize results, rather than merely break even. When an oil company evaluates its options, for example, it has to keep track of the price of a wide range of refinery products, from gasoline to wax to plastic feedstocks, and adjust the refinery output to maximize its revenues.

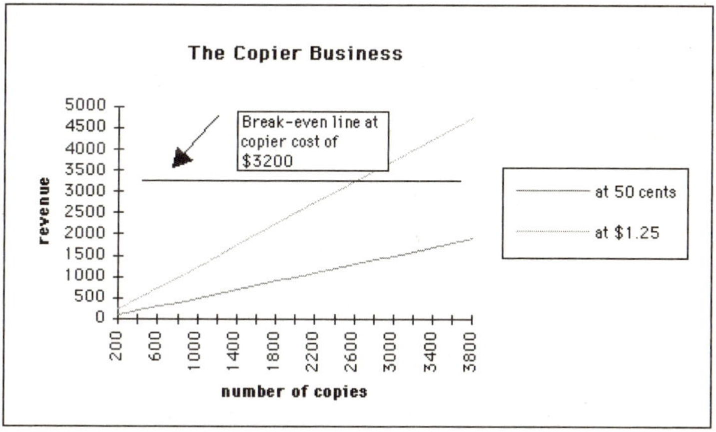

Figure 10-1: The break-even problem.

Paying off

Producing a formula for payments is a pretty complicated business, because interest is being charged while you are making payments and the amount of interest versus principal is changing with each payment. The simplest way to solve any problem involving time payments is to use a financial calculator, the second simplest is to use the functions in a computer spreadsheet, and the hardest way that actually still works is to use a scientific calculator with a y^x function key.

The payment formula for mortgages and ordinary consumer loans looks like this:

$$PMT + PV \times \frac{i}{1-(1+i)^{-n}}$$

For a bit of later-in-life algebra, you can try plugging in some numbers. I'm going to use the example of a car payment. I'll try a car that costs $12,500 and a payment plan that goes to 4 years at 9.1 percent interest.

The symbol *PV* in the formula stands for *principal value,* which is just the loan amount, so

$$PV = 12{,}500$$

n is the number of periods in the loan. It could have been 4 for 4 years, but interest on these consumer loans is always computed monthly, so

$$n = 4 \times 12 = 48$$

because there are 48 months in the 4 years.

i is the interest per period. That means that the number to plug into the formula for *i* is

$$i = {}^{0.091}\!/_{12}$$

$$= 0.00758333$$

That is, 9.1 percent is the same thing as the number 0.091, and you have to divide it by 12. Also, you have to keep a ridiculous number of decimal places in interest, since this interest number gets multiplied by itself many times and errors can start to accumulate.

Putting the pieces together, you get

$$(1 + i) = 1.00758333$$

Calculating $(1 + i)^{(-n)}$ is the exciting part. It's

$$(1 + 0.00758333)^{-48}$$

You enter the number 1.00758333 in the calculator, press the y^x key, enter –48 (which you do by entering 48 and then pressing the +/– key), and press the = key.

That complicated factor works out to

$$(1 + 0.00758333)^{-48} = 0.69584621$$

Plugging in the terms, you have

$$\begin{aligned} PMT &= 12{,}500 \times (0.00758333)/(1 - 0.69584621) \\ &= (94.791666)/(0.30415790) \\ &= 311.65 \end{aligned}$$

Run through this kind of calculation a few times in an afternoon, and you'll be running down to K-Mart for a basic financial calculator that just has keys that say *i, n, PMT,* and *PV.* If you're clever and organize your work carefully, you can avoid entering these long numbers more than once, but if you're not that clever, it's really a pain.

Anyway, this exact formula is programmed into calculators and spreadsheets, and if you have lots of patience, you can derive it from simple considerations about interest compounding.

Geometry and Reality

Euclidean logical proof meets with little application in daily life, mainly because so few real-world axioms can be trusted absolutely. On the other hand, the actual physical world is built from the formulas of arc length, diameters, areas, and volumes that everyone learns in geometry class.

The gritty details

EXAMPLE

Suppose you have a space near the back door of your house and want to put in a concrete area that's 9 feet x 12 feet. This area will be a great assistance to fire safety when you use whole quarts of lighter fluid to ignite mere handfuls of charcoal briquettes for summer barbecues. Consider the geometry problems involved:

You need to construct a wooden frame for pouring the cement. For a nice, thick slab, you may want to use 2 x 8 boards. Of course, following the time-honored customs of building-material practice, a 2 x 8 is actually somewhat smaller than 2 inches x 8 inches in cross-section. You are making a frame that's 9 feet x 12 feet, so you need to buy the amount of lumber shown in Figure 10-2. That gives you the perimeter of the frame, 9 + 9 + 12 + 12 = 42 feet.

Now you need to figure out how many bags of cement to buy. The diagram in Figure 10-3 sets up the calculation. One of the first things you see (and when you think about the volumes involved, it makes sense) is that having a pickup truck to lug all that cement back from the hardware store would be a great convenience.

Now suppose you got really artsy and decided to try for the fancier shape shown in Figure 10-4. Hey, people do stuff like this all the time. Again, you would be calling on old geometry rules, and with a little bit of attention, you would notice that you are adding the volume of a very shallow cylinder to that of a very shallow rectangle.

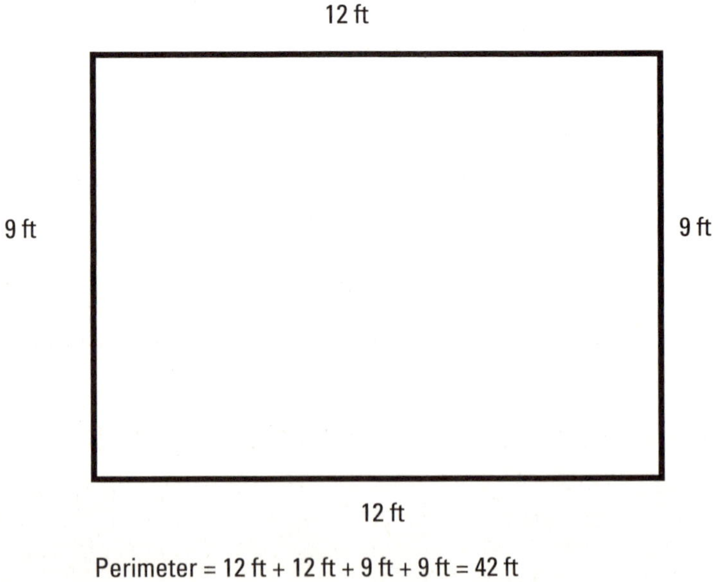

Figure 10-2: A perimeter question.

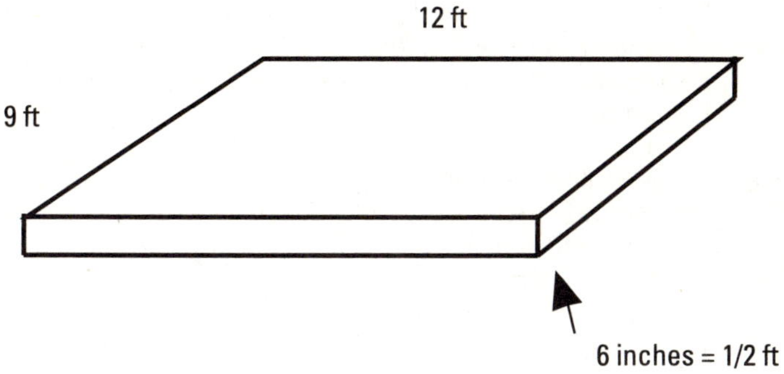

Volume of cement is 12 ft x 9 ft x 1/2 ft = 54 cubic feet

Figure 10-3: Cement volume.

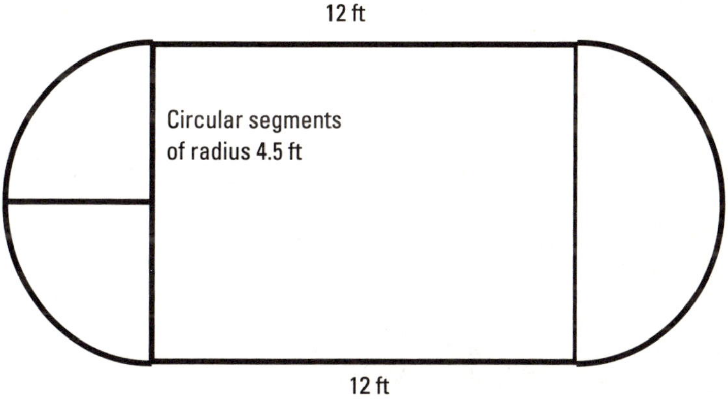

Volume = volume of 9 ft x 12 ft x 1/2 ft slab

+ volume of cylinder of radius 4.5 ft x 1/2 ft height

Figure 10-4: Cement volume, the artistic version.

A world of impossibilities

One of the most interesting features of geometry, although it may not be discussed extensively in school, is that outside the world of direct measurement, geometry generates problems of mind-boggling difficulty.

For example, once you try to cover an area with tiles that aren't rectangles or squares, you find thorny problems all over the place. Can you figure out how to cover a flat surface with the kind of tiles shown in Figure 10-5?

The simplest problems turn out to be very challenging. If you fill a cylinder with spheres, what fraction of the space in the cylinder will be occupied? If you fill a rectangular box with little cylinders, is that really the same problem as the sphere problem? Some of the problems in this class have been solved, while others, which you could easily state to a third-grade class, are still considered research problems. It's satisfying, in a way, that geometry generates so many interesting problems, often with no apparent business or practical applications. Euclid would be delighted.

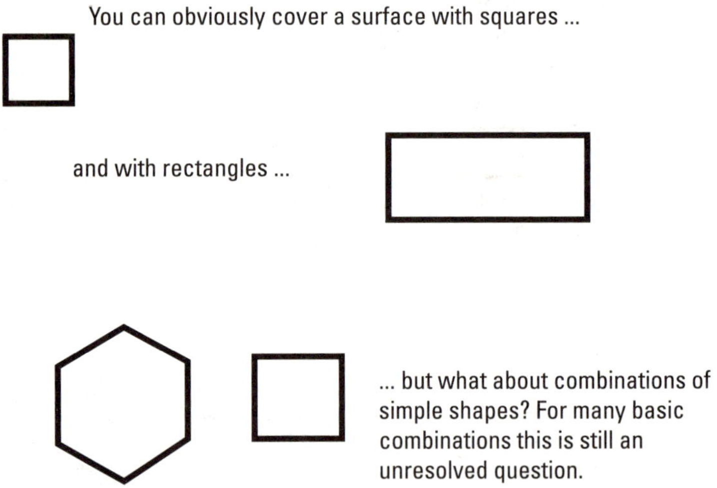

Figure 10-5: Exotic tiles in geometry.

Thinking about Risk

Simple considerations of numbers and their relative sizes connect directly with daily life — in fact, that's one of the main themes of this book. Every day you are bombarded with information presented in terms of numbers, and if you think about the numbers a bit, you can decode the reality behind them. Often, this ability is crucial, as much of the information we get is distorted both by its presentation and by our own internal perceptions. When you make business decisions — for example, decisions about insurance — you make better decisions if you start with numbers that reflect reality.

For example, a perusal of the news on a daily basis might convince you that you are at immediate risk of death from murder, floods, riots, or accidents. You almost never have the opportunity to step back from the flow of news and ask yourself how any of the odds in these situations affect you. (See Chapter 5 for more insight into statistics in the news.)

Table 10-1 gives you a little homework table to photocopy. See if you can fill in the blanks over a period of weeks by paying attention to newspaper articles. Particularize the results to your own state or even to your own city if you can get the data. National statistics, for example, don't mean much — one set of numbers is appropriate for you if you live in Boise, and quite another set works for Houston.

Table 10-1 Your Own Risk Directory

See if you can fill in the blanks from newspaper and magazine reading. One blank has been filled in to give you some starting information.

Event	*Risk*
Serious car accident	
Commercial airplane accident	1 in a million per 10 hours
Cancer (lifetime)	
Heart attack	
Skydiving (per event)	
Murder by stranger	
Home accident (all kinds)	
Natural disaster	

TIP

Looking over the numbers, you may find that you have been worried about getting killed in a carjacking, whereas your real problem is that you smoke two packs a day. Practically everyone in California, for example, is at thousands of times more risk from drunk drivers than from earthquakes, despite rampant earthquake paranoia. Can you determine the difference in the chances of being shot in the best and in the worst neighborhood in your area? Should you relocate your business, or is your current location safe enough? Take it from me: The numbers are worth collecting. Even if you don't use it much, a perspective on numerical reality is a helpful thing to keep on hand.

Chapter 11

Ten (or So) Numbers to Remember

In This Chapter

- Numbers you need to know
- Numbers to know for fun

Okay, you don't have to remember all of the numbers in this chapter. Some are useful, and some are just for party chit-chat or to prove that you have an education. What is remarkable is the way individual numbers can have long stories, historical associations, and odd properties.

Pi (π)

Recall my discussion of geometry, if you will. Pi, or π, is the ratio of the distance around the edge of the circle (the circumference) to the distance across the middle (the diameter). In other words,

$$\pi = \text{circumference/diameter}$$

Therefore

$$\text{circumference} = \pi \times \text{diameter}$$

Although you are taught in school that $^{22}/_7$ (or $3 + {}^1/_7$) is a "good enough" approximation to use, just the number 3 isn't so bad if you have trouble remembering fractions (3 appears to be the value recommended in the Old Testament). If you have a circle that's 1 foot across, using 3 for π gives you circumference = 3 × 12 inches (1 foot) = 36 inches (1 yard)

Using the eighth-grade school value of $22/7$, you get

circumference = $22/7$ × 12 inches = 37.7143 inches

The exact circumference (using a better approximation for π) would be 37.6992 inches (to just four decimal places). For a geometry adventure of this size, the difference between $22/7$ and real π is 0.015 inches — less than a sixtieth of an inch. That's actually less than the width of the ink on the lines of a yardstick.

One common value of π in Egyptian practice was 3.16, which, since it's correct to 1 percent, probably didn't produce any errors worth arguing over on the banks of the Nile. The first really serious calculation of π was probably due to Archimedes, who used many-sided regular polygons to make an estimate of approximately 3.1418. By later times in the Greek world, the number 3.1416 was in common use.

In China, the number 3.162..., or the square root of 10, was a popular choice for π. It was recognized that the real value was closer to 3.14, but because 10 is a "perfect" number with lots of symbolism, the Chinese had a romantic attachment to its square root. By 400 A.D. or so, Chinese astronomers had worked out $355/113$ as an approximation. This fraction is 3.1415929204; when you compare it to the real value of 3.1415926535..., you may realize that it would be very difficult to determine this number by measurement since it's correct to a few parts per million and no one in fourth-century China had instruments of that accuracy. How was this done? Beats me.

Avogadro's Number

This number connects the world of individual atoms with the world of amounts of materials you can actually see. For short, the value is just 6.02×10^{23}.

Here's a rough explanation. Hydrogen is the lightest element, with an atomic weight of 1. Oxygen has an atomic weight of 16. A water molecule (H_2O) has 2 hydrogens and 1 oxygen, for a molecular weight of 18 (16 + 1 + 1 = 18). A molecular weight of 18, in turn, means that 18 grams of water is 6.02×10^{23},

Avogadro's number, of water molecules. This quantity is called a *mole*.

The main point here is that, because atoms are very little, this number is very big. There are more silicon atoms in a grain of sand than there are grains of sand on a beach. Just for fun, prove this for yourself by picking some arbitrary number for grains per cubic inch and number of cubic inches in a good-sized beach.

If a grain is $\frac{1}{50}$ of an inch on a side, you can figure that there are $50 \times 50 \times 50$ grains in a cubic inch. Then you can figure the number of grains in a chunk of beach a half mile long by 50 yards wide by 20 feet deep. In inches, that's

$$(12 \times 5{,}280 \times \tfrac{1}{2}) \times (12 \times 3 \times 50) \times (12 \times 20)$$

That works out to about 1.7×10^{15} grains. One grain of sand is much less than a mole of silicon dioxide (I'm assuming plain old sand, not cool black sand from special lava or something like that). If you figure that it takes approximately 0.5 million of the little grains to make up a mole of silicon dioxide, then

$$\text{Silicon atoms in a grain} = \frac{(6.02 \times 10^{23})}{500{,}000} = 1.2 \times 10^{18}$$

So really, there are about a thousand times as many silicon atoms in a grain of sand than there are grains in a decent beach (I'm thinking specifically of Moonlight Beach in Encinitas, California, and would welcome funding for experimental testing of this proposition from July to September — you pick the year). In defense of beaches everywhere, I should point out that Copacabana is close to 5 miles long and almost 500 yards wide, making it a better numerical competitor with the single grain. But 20 feet is actually pretty deep for beach sand, so the grain still wins by a bit.

1/1,000

In the ideal world of Greek mathematics, numbers existed in a pure, theoretical realm. They could be calculated, in principle, to any degree of accuracy, but they existed as some perfect form in the eternal mind of the universe. Now that computers can do the arithmetic, performing all sorts of calculations with arbitrary precision is relatively easy. Want π to 40 decimal places? No problem at all.

The issue of precision is one point in which mathematics departs from the real world of stuff. In the real world, it's often difficult or expensive to get a measurement that's accurate to one part per thousand. Consider a few examples:

- **Rulers:** Head on down to a stationery store and find the flexible plastic rulers in the school supplies section. Pick out six of them and line them all up at the left-hand edge of the ruled part. Now look at the right-hand edge, where the rulers indicate 12 inches. Usually, there's a clearly visible discrepancy — after all, the process of making these rulers just involves printing on fairly soft plastic. Do you think that a plastic ruler can be accurate to one part per thousand (that is, one hundredth of an inch)? Heat one up in warm water and repeat the comparison if you do!
- **Lego blocks:** If plastic objects have such sloppy dimensions, how can Lego make blocks that can be fitted into smooth walls hundreds of blocks high? Well, Lego needs to hold block tolerances of one part per ten thousand, and that means that Lego has to design its own plastic injection molds using its own special technology. As a result, the company produces toys that have greater precision than most plastic aerospace or medical parts.
- **Thermometers:** You used to be told that 98.6 degrees Fahrenheit was the absolute normal temperature. So that you could find out whether you were normal, classic mercury-based thermometers were developed to display only a few degrees over a measuring span of several inches. That made a tenth of a degree correspond to about a tenth of an inch, producing very high accuracy indeed.

  Lately, measurements on tens of thousands of people have shown that the 98.6 number, faithfully and religiously repeated through decades in textbooks, simply isn't an absolute standard. Plenty of people have temperatures as low as 97 day in and day out, and plenty of other people hover between 99 and 100 degrees without having a fever.

- **Volume:** Special chemical laboratory glassware is calibrated to be accurate to about one part per ten thousand, as long as the volume is large enough. Kitchen glassware — a measuring cup, for example — is accurate to about one percent. These days, computer-controlled gasoline pumps report your gasoline purchase to a thousandth of a gallon, although a spot check by the state of California found the typical pump to be off by a tenth of a gallon or so. In other words, the pumps give you plenty of digits, half of which may as well be random.

TIP

The reason I think that $\frac{1}{1,000}$ is an important number is that it's the working definition of "good enough." If you know how much money you have to 0.1 percent accuracy, that's one part per thousand accuracy and should be good enough. One part per thousand means about one-fourteenth of an inch on a two-yard piece of cloth from a material store, and it's about the width of a line of grout on a tile floor in a standard-sized room. Keep this in mind when you look at results on a calculator — here in the real world, you can usually relate only the first three digits to physical stuff.

MCMLXXIV

It took an amazingly long time — hundreds of years — for the number system pioneered in India to work its way to Arabia and then across the Arab world to Spain, from which it was finally introduced to northern Europe. Contemporary reports suggest that the concept of zero was mind-boggling enough to give people headaches. These days we use "Arabic" numerals like 1, 2, and 3 without a second thought, and even kindergartners know about zero.

So except among mathematical hipsters, Roman numerals (which don't include a zero) persisted well past the year 1000 A.D. They still appear in movie credits, on medical school diplomas, and in several other areas where a little artificial augmentation of prestige is thought to be helpful. Decoding this particular number will at least help you with any number in this century.

You read it like so:

- M = 1,000.
- CM = 100 before 1,000, which means 1,000 – 100 in Roman. When a number is "out of order" (C is one hundred, and it's out of order because it appears before the M), you subtract it from the bigger number to its right. So CM is 900.
- L = 50.
- XX = 20. X is 10, XX is 20, and XXX is 30. If you had thought in terms of this very concrete kind of counting all your life, the notion of zero probably would have struck you as pretty spacey, too, just as it did the medieval monks who served as custodians of arithmetic.
- IV = 4. I is 1, II is 2, and III is 3. What more could you want? At 4, there's a little confusion. Sometimes it appears as IIII, the obvious choice, but sometimes it's IV, which is 4 according to the out-of-order rule since V is 5.
- The date MCMLXXIV is thus 1,000 + 900 + 50 + 20 + 4 = 1974.

The Romans, by the way, typically used an interesting design of abacus to do arithmetic, avoiding working out problems on paper. Using their humble abacuses and lots of common sense, they could solve most engineering problems. The number system wasn't very streamlined, but Roman engineers had a vast repertoire of tricks and work-arounds for calculations. As nearly as archaeologists can reconstruct the situation from Pompeii and from Rome's port city of Ostia, the Romans around 100 A.D. achieved a standard of living, at least in some cities, that was not matched again until around 1900.

$\frac{9}{5}$, 32, and $\frac{5}{9}$

Welcome to the U.S., the last place on earth to rationalize its measurements. Really, it's something of a miracle that we ended up with a decimal-based money system, approved in a short-lived fit of anti-British sentiment. I won't go on about inches and furlongs and quarts and pecks, but I must make some remarks about temperature. That a large chunk of elementary school science class has to be devoted to teaching the way scientists and the entire rest of the planet measure things is a pretty strange feature of American civilization.

We still use Fahrenheit temperatures, and everyone else (here called EE) uses the Celsius (also known as centigrade) scale. Both scales are based on the freezing and boiling points of water. On the Fahrenheit scale, 32° = freezing and 212° = boiling. Hey, what else?

On the Celsius scale, 0° = freezing and 100° = boiling, which in some sense certainly looks more promising. To get back and forth between the two scales, here's what you do.

Converting EE to U.S.

Their numbers are smaller. On a crummy, rainy spring day in Paris, the weather news may report a temperature of 5 degrees. The rule is: First you have to change the scale, and then you have to add a step.

Change the scale by multiplying the Paris number by $\frac{9}{5}$. That gets you

$5 \times \frac{9}{5} = 9$ degrees

Then you add the step, the 32-degree offset that Herr Fahrenheit built into his thermometers. That gets you to

$9 + 32 = 41$ degrees

A rainy 41 degrees in Paris — sounds like Armagnac weather to me, kids.

By the way, if you don't feel like multiplying by $\frac{9}{5}$ in your head, just multiply by 2 instead. You end up with a number that's higher by a few degrees, but hey, it's just weather. Personally, I can't tell 41 from 43 degrees, anyway.

Converting U.S. to EE

The reverse process is this: Take off the 32-degree step and then change the scale.

Suppose you're trying to explain to an Italian why most of the American Southwest is uninhabitable without air-conditioning (some would say that it's still uninhabitable from a long-term ecological water-management viewpoint). You are making your point by discussing the week in Gila Bend, Arizona, where the temperature went past 122 degrees every day for a week.

First you step the number down by 32, so that 122 becomes

$$122 - 32 = 90$$

Then you change the scale. Remember that EE numbers are smaller than ours, so you multiply by $\frac{5}{9}$. That gives you 50 degrees. The last time your Italian friend saw that number on a weather map, it referred to some spot in Libya about 100 kilometers inland from the Mediterranean coast. Sounds like Campari and soda weather to him, assuming that he'd go near the place.

Again, if you don't like all this $\frac{5}{9}$, $\frac{9}{5}$ business, you can just divide by a factor of 2. In this example, you get 45 degrees Celsius. That still sounds like a blazing inferno to the EE community.

Megabytes

A whole new set of popular numbers has appeared in the wake of the computer revolution. Bits, bytes, and so forth are standard terms for measuring computer memory.

A bit is either a zero or one and can contain as much information as the state of an on/off switch — it's a one or a zero, so it represents on or off electronically.

A byte is a number composed of eight bits. Each bit can represent two states. That means that a byte can take any value (represent any number of states) between one and the number

$$2 \times 2 \times 2 \times 2 \times 2 \times 2 \times 2 \times 2 = 2^8 = 256$$

although, in computers, a byte is used to represent the numbers from 0 to 255.

That's a pretty good range, because it can represent all the characters and numbers typically used in printed English. English and European languages, in the world of hard-core computer designers, are called single-byte languages because they can be represented with a single-byte alphabet. Because they have much larger character sets, Japanese and Chinese are called double-byte languages because two bytes can represent 65,536 characters. That's more than enough to look up the 40,000 characters used in Chinese and Japanese books in a computer-based lookup table. So you can assign a two-byte number to each character in a Japanese dictionary and thus represent Japanese (with some difficulty) inside a computer.

A megabyte is approximately a million bytes. More exactly, it equals

$$2^{20} = 1{,}048{,}576$$

Back in the late 1980s, a computer with 2MB of memory and a 200MB hard drive was considered pretty fancy. Now I'm constantly asked by people at my computer-magazine job if such a computer is even good enough to do elementary school homework. It's a very odd business, really — the computer industry has convinced everyone that computers that would have been the Big Computers at NASA a decade ago aren't fit for kindergartners. This attitude has more to do with marketing than reality.

666

This is the number of the Beast in the Book of Revelations, and it's a great starting place for mystical speculations. The game here is to decide who the Beast is by assigning numbers to letters in a name and then adding up the letters to get 666. It is widely believed that the first version of this was meant to refer to the full Latin name of the emperor Nero.

By the time of the Reformation, Catholic scholars had determined that one spelling of the name of Martin Luther worked out to 666, and Protestant scholars returned the favor by finding 666 in the name of Pope Leo X. As you might expect, numerologists have associated the number with Adolf Hitler, Franklin Roosevelt, and a host of less likely Beast celebrities, on down to Rush Limbaugh and Microsoft's Bill Gates.

One conventional assignment for this kind of calculation is a = 1, b = 2, up to z = 26. Sometimes you have to assign a = 26 and so forth instead of a = 1 to make things work, and sometimes you have to cross out the letters that aren't used in Roman numerals, but of course it's well worth the effort.

1,729

While in a hospital in England, Indian mathematician Srinivasa Ramanujan was visited by his coworker G. H. Hardy, one of the leading mathematicians in England in the first part of this century. Hardy remarked that he had arrived in taxi cab number 1,729, said that it seemed like an uninteresting number, and hoped that it wasn't a bad omen. Ramanujan said, "No, it's a very interesting number. It's the smallest number that can be expressed as the sum of cubes in two different ways." That is, he had figured out in a fraction of a second that

$$1{,}729 = 13 + 123 = 103 + 93$$

The trick was not just to have seen this but also to have figured that there are also no numbers smaller than 1,729 for which you can make a similar decomposition. And hey, the guy was lying there with the flu! I should also emphasize that Ramanujan's contributions were all serious, almost weirdly original mathematics in infinite series, continued fractions, and related topics.

The Golden Ratio

Greek mathematicians concluded that a rectangle in which the long side is 1.618 times longer than the short side is particularly aesthetically appealing. This ratio appears in the construction of Greek temples, not just in the overall proportions but in the shapes of doorways and windows. Although it's not clear that the Egyptians had the same mystical associations with this number that fascinated the Greeks, the number also appears in pyramid construction. A large-scale 19th century study of all sorts of rectangular shapes (bricks, playing cards, greeting cards, and so forth) from cultures all over the world concluded that people have an intuitive preference for this kind of rectangle compared to rectangles that are either shorter or thinner. Shockingly enough, legal-size paper is 1.64 times as long as it is wide — that's close enough!

The exact number derived by the Greeks is

$$\Phi = \frac{\left(\sqrt{5}+1\right)}{2}$$

$$\tilde{}$$

$$= 1.61803\ldots$$

So 1.618 is certainly a close enough approximation if you want to make your own Greek temple.

10^{14}

Just a year ago, the astrophysics community generally agreed that the universe was 10^{14} years old. Then things got strange. First, using standard techniques but new information from space-based telescopes, astronomers found objects that seemed to be older than the "established" age of the universe. And then they observed objects that appeared to be farther away than any yet observed but that were ridiculously young by astrophysical standards. According to the pretty-much accepted Big Bang theory, you are supposed to see the old stuff, not new stars, at the outer edges of the universe.

Reasonable scientific evidence (setting aside for the moment possible religious objections) shows that the Earth is about 4×10^9 years old, so it's only been around for a blip in terms of the history of the universe. It's the rest of the sideshow where the dating is more or less out of control at this point. I put this number in the collection because, unlike π or 666 or 1,729, you'll see it toggled up and down in newspaper accounts of the latest theories in cosmology. Please feel free to update it every six months or so in your own copy of this book.

Chapter 12

Ten Fast, Silly Number Tricks

In This Chapter

- The age trick
- The birthday trick
- A kid trick
- Grain and the pharaohs
- Almost a puzzle
- Fiscal humor
- A dicey business
- Mental miracles
- The Afghani birthday problem
- More fabulous powers

It's highly appropriate that there's a chapter like this in this book, even though as a professional grump I of course disapprove of all frivolity. After all, who actually says (beside you and your friends) that math isn't fun?

The Age Trick

This goofy party trick assumes that your friends can at least add and subtract a bit. To do the age trick, follow these steps:

1. **Ask someone to multiply his or her age by 10.**

   Suppose Steve's age is 25. He multiplies it by 10 and gets 250.

2. **Have that person subtract 9 times any number from 1 through 9 and ask for the difference.**

   Steve subtracts 45 (5 × 9) from 250 to get 205.

3. **Add the last digit of the number to the first two to get the age.**

   You hear the number 205 and partition it into 20 and 5 and add the two numbers together, giving you 25.

Suppose Steve had picked some other multiple of 9. Just to be complicated, suppose he picked 81 (9 × 9). That gives him

$$250 - 81 = 169$$

as the key number. You separate it into 16 and 9, add them together, and announce the result: 25!

What's going on here? Actually, the trick is very similar to the method of "casting out nines" to check on any addition or subtraction. Note that the trick works with *two*-digit ages — it fails for six-year-olds and for people over 100. On the other hand, people under ten and over 100 are very rarely reluctant to state their ages. I once went to a party where I was persistently reminded by a young lady that she had just turned four a few days earlier, and I was accosted several times by an elderly gentleman who reminded me, shaking his head as if in disbelief, that he was more than a century old. "More than a century . . . can you imagine that, Sonny?" The guy is now 105!

The Birthday Problem

The birthday problem is a standard in probability theory, but it has wide application. The question is how many people do you have to have in a room before it becomes likely that two people have the same birthday? Now, please note that this is very different from trying to calculate the odds on someone showing up with a particular given birthday, say April 26. Here's how you work this problem:

1. You have one person in the room. A second person walks in. The second person has a 364/365 chance (0.997 — that's a 99.7 percent chance) of not having the same birthday.

2. Now a third person walks in. This person has to miss two days of the year to not have the same birthday as either of the others. The odds on that are 363 in 365 (0.995).

3. Now a fourth person walks in. This person's birthday has to miss three other days of the year. Those odds are 362 in 365 (0.992). That is, the person has 362 chances out of the 365 days to miss the other birthdays.

4. The chances of everyone not having the same birthday can just be calculated as

   Probability that all birthdays miss each other =

   $(364/365) \times (363/365) \times (362/365) \times (361/365) \times (360/365) \times (359/365) \times \ldots$

   $= (0.997) \times (0.995) \times (0.992) \times \ldots$

   $= (0.989) \times (0.986) \times (0.984) \times \ldots$

   All these numbers are pretty close to 1, but you can convince yourself that if you multiply enough numbers like 0.96 together, you rapidly get a pretty small number. Get out your calculator and multiply 0.96 by itself ten times!

5. The big product of all these 0.99-type numbers falls to less than 0.5 when you multiply 26 of them together. To be on the safe side (building a few more numbers into the factor), you can predict that two people will have the same birthday when there are 30 or so of them in a room.

People find this result surprising, but then coincidences catch people off guard. You find yourself seated on an airplane next to someone who was in your American History class in tenth grade, and both of you are knocked out by the coincidence. But the chances of something like this never happening to you are actually fairly small. As in the birthday problem, you are saying that some coincidence is likely, but you don't know which coincidence it will be (you don't know what the magic birthday will be).

Pennies from Heaven

That last section was actually an outburst of real math, albeit applied to a nonlethal problem.

Ask someone this question (preferably someone under the age of nine):

Why are 1995 pennies, in good condition, worth almost $20?

The answer is that 1,995 pennies, all piled up, are only 5 cents short of being 2,000 pennies. This is almost the numerical equivalent of a bad pun.

The Latest from 1600 B.C.

If each of seven houses has seven cats, and the seven cats are each served by seven mice, and each of the seven mice is hoarding seven ears of grain, and each grain has seven kernels, how many kernels of grain do you have?

$$7 \times 7 \times 7 \times 7 \times 7 \times 7 \times 7 = 823{,}543$$

I just threw this in here because of its outstanding formal resemblance to the nursery rhyme: "As I was going to St. Ives, I met a man with seven wives. . . ." It comes from the Rhind Papyrus, an Egyptian mathematics text for training scribes.

Switching Digits

Figure out which two digits need to be switched to make this multiplication correct:

$$\begin{array}{r} 324 \\ \times\ 68 \\ \hline 2992 \\ 2744 \\ \hline 25{,}432 \end{array}$$

You can write this down on a scrap of paper and leave it around an office for someone else to find. It has been used to drive people crazy for several decades. Actually, problems of this type are fairly easy to construct if you want to make a career out of being obnoxious.

Money, Money

Offer to pay a hundred dollars to anyone who can give you five dollars in ten coins. The coins have to be half-dollars, quarters, and dimes, and you have to use at least one of each. Amazingly, people will spend several minutes counting change and making little calculations on this, and some never give up.

If the person could give you ten half-dollars, that would be $5. The requirement that you use other coins makes it impossible.

Dice Trick

This is essentially a math trick because it's based on addition and subtraction. You hand someone three dice and tell him or her to stack them up. There are now five hidden faces, meaning the parts of the dice that can't be seen. You ask your friend to check the total of the dots on the hidden faces while your back is turned.

When you turn around, you simply notice the number of dots on the face on the top of the stack. After much grimacing and mock concentration, you mentally subtract this number from 21 and announce the result.

The story is that the number of dots on the opposite faces of each dice add up to seven. For three dice, the total number of dots on the opposite faces is 21. In this case, the top face is showing, so you subtract the number of dots on the top face from 21.

Speed Arithmetic

Pick two single-digit numbers. For this example, use 3 and 7. Then write them like so, and make a series of numbers underneath them by adding them in pairs until you have ten numbers. It looks like this:

- 3
- 7
- 10
- 17
- 27
- 44
- 71
- 115
- 186
- 301

Because you have read *Math Basics For Dummies,* you can now do fabulous calculations in your head. You claim that you can add these numbers in your head at a glance. The trick here is to multiply the fourth number from the bottom by 11. It's 71, and 71 times 11 is 781 ($71 \times 11 = 71 \times 10 + 71 = 781$).

You may have noticed that except for a few tricks in the previous section, there isn't much mental arithmetic in this book. Looking over the earlier literature on mental arithmetic from 1900 to the present and evaluating it, I concluded that a trick for rapidly multiplying large numbers by 17 (of course there's a special trick for it) is not especially helpful in daily life. If I could have you remember just a few things a year from now, I would rather that it be how to make compound interest work for you rather than against you. Even if you never learn tricks for doing long division in your head, you should have a bad gut feeling about financing ski vacations on 22 percent credit cards.

Afghani Birthdays

This is a variation on the birthday problem discussed earlier. You are at a party in a large city, and some of the people at the party are from Pakistan, Afghanistan, Iran, and other Islamic countries.

You propose that you cannot only see that some of the people have the same birthday, but you propose that you know the birthday. It's January 1! Wow, all these people are Capricorns!

What happens is that in many societies, people just don't record actual birthdays, especially in small rural villages. When people from those societies get passports to come to the United States, the birthday is usually listed as January 1. In the hardball version of this trick, you size up the party and tell some unassuming American friend that it's a mathematical certainty that at least two people at the party have a birthday on January 1. The people with the arbitrarily dated passports are, of course, automatically in on the joke because they know that the date was assigned in an office.

Our focus on birthdays in the West probably comes from the association of the calendar with saints' days and the practice through most of European history of naming Christian babies after saints. Actually, from a practical point of view, knowing your own age to the exact day becomes pretty irrelevant past the age of seven or so, unless you live in a society like ours that sets great store by exact dates.

The All-Seeing Mind

This is a calculator trick. Have someone perform these steps on a calculator, and claim to be following mentally:

1. **Enter any three-digit number (the fabulous secret number).**
2. **Multiply the number by 10.**
3. **Subtract the secret number.**

4. **Divide this result by the secret number.**
5. **Square this number.**
6. **Add 19.**

After appropriate theatrics, you grandly announce the result: 100. What's really going on here is that the first four steps simply turn any number into 9 (effectively, you are dividing the number into 9 times itself). The square of 9 is 81, plus 19 is 100. But if you have a victim for this trick who's truly hopeless at arithmetic, he or she won't see the gimmick.

Index

• Numbers and Symbols •

• A •

• B •

• C •

• D •

• E •

• M •

• N •

• O •

• P •

• Q •

• T •

• U •

• V •

• W •

• X •

• Y •

• Z •

Your fun and easy formula
for understanding basic chemistry
Chemistry
FOR
DUMMIES
PORTABLE EDITION
John T. Moore
Associate Professor of Chemistry, Stephen F. Austin State University
A Reference for the Rest of Us!
Fun and easy guide
that brings biology to life!
Biology
FOR
DUMMIES
PORTABLE EDITION
Donna Rae Siegfried
Author of Anatomy & Physiology For Dummies
A Reference for the Rest of Us!
FREE eTips at
dummies.com

The fun and easy way® to speak and write with confidence
English Grammar
FOR DUMMIES®
PORTABLE EDITION
Parts of Speech
adjective - describes somet
somebody. ex.→lumpy, soft, ug
adverb - tells how somethin
It modifies a verb and usu
FREE eTips at dummies.com®
ence for the Rest of Us!®
The fun and easy way® to start speaking Spanish
Spanish
FOR DUMMIES®
PORTABLE EDITION
Words to Know
mucho moo-cho a lot; much
difícil dee-fee-seel difficult; hard
fácil fah-seel easy
gustar goos-tahr to like
hablar ah-bvlahr to speak; to talk
este ehs-the this one
otro oh-troh the other one
Susana Wald
Language instructor and translator
A Reference for the Rest of Us!®
FREE eTips at dummies.com®